D0772554

THE SCIENCE OF CHEESE

The Science of Cheese

Michael H. Tunick

OXFORD
UNIVERSITY PRESS

OXFORD
UNIVERSITY PRESS

Oxford University Press is a department of the University of Oxford.
It furthers the University's objective of excellence in research,
scholarship, and education by publishing worldwide.

Oxford New York
Auckland Cape Town Dar es Salaam Hong Kong Karachi
Kuala Lumpur Madrid Melbourne Mexico City Nairobi
New Delhi Shanghai Taipei Toronto

With offices in
Argentina Austria Brazil Chile Czech Republic France Greece
Guatemala Hungary Italy Japan Poland Portugal Singapore
South Korea Switzerland Thailand Turkey Ukraine Vietnam

Published in the United States of America by
Oxford University Press
198 Madison Avenue, New York, N.Y. 10016

Library of Congress Cataloging-in-Publication Data
Tunick, Michael.
The science of cheese / Michael H. Tunick.
pages cm.
Includes bibliographical references and index.
ISBN 978-0-19-992230-7 (alk. paper)
1. Cheesemaking. I. Title.
SF271.T86 2014
637'.3—dc23 2013010729

9 8 7 6 5 4 3 2 1

Printed in the United States of America
on acid-free paper

For Gail, Dan, and Susan

Contents

Preface

> Cheese. The adult form of milk. Hard and soft; new, mature, and overripe.
> Graced with a spectrum of smells from wispiness to shocking assault upon the nostrils.
> —RICHARD CONDON, *A Talent for Loving*, 1961

PEOPLE HAVE BEEN enamored with cheese for a long time. The intricate combinations of appearance, aroma, flavor, and texture have inspired writers to refer to cheese as "milk's leap toward immortality" (Clifton Fadiman), "the soul of the soil" (Pierre Androuët), and "the wine of foods" (Vivienne Marquis and Patricia Haskell). Cheese has much in common with wine and beer: they result from fermentation by microorganisms, they are "value-added" products (processing greatly increases the value of milk, grain, and grapes), and they reflect the local climate and terrain. Cheese may be kept for months, and traditionally provided a source of protein, vitamins, and minerals during the winter when other food was scarce. Cheese is nutritious and delicious, and can be appealing or appalling. There has been a great deal of research on this complicated food—over 20,000 scientific papers about some aspect of cheese have been published over the past ten years alone. Some authors have distilled this information down to the basics for the general public, and others have written works that most people without a scientific background are unable to understand. I've tried to take the middle ground, by writing about the science without getting too technical. Like many books on this subject, we will cover milk (Chapter 1), the fundamentals of making cheese (Chapter 2), aging cheese (Chapter 3), and the different classes of cheese (chapters 4–13). But we will also cover the chemical compounds involved and how the flavors arise as we talk about cheese classes, and we will detail nutrition (Chapter 14), how cheese is analyzed (Chapter 15), rules, regulations, and specially designated cheeses (Chapter 16), things you can do at home (Chapter 17), and other topics besides.

This book is a product of the accomplishments of two organizations, the American Chemical Society (ACS) and the Agricultural Research Service (ARS). ACS has over 160,000 members worldwide who conduct research in chemistry, teach it, or write about it. Scientific societies like ACS promote knowledge by holding meetings where scientists present results of their research, by publishing journals and books based on these results, and by providing online information, including webinars. ACS began offering webinars to its members in 2010 and started a series called "Chemistry of Food" the following year. The first presentation was given by Charles Bamforth on the chemistry of beer, and I had the third talk, titled "The Chemistry of Cheese and Why We Love It." Oxford University Press had already published three editions of Charles's book *Beer: Tap into the Art and Science of Brewing*, and they were looking for someone to write a general-interest book on cheese chemistry. They viewed my webinar, invited me, and the result is in front of you.

ACS has over 30 technical divisions dealing with various areas of chemistry. One of these, the Division of Agricultural and Food Chemistry (AGFD), has been an all-volunteer group since its formation in 1908. I was chair of AGFD in 2001 and am its secretary now. Most divisions including AGFD hold symposia at the two annual ACS meetings in March/April and August/September. Some of our recent symposia have covered such diverse topics as nanotechnology, tree nuts, bioactives, cereal grains, waste utilization, subtropical fruits, consumer safety, chocolate, allergens, Hispanic foods, flavors, and many more. ACS and AGFD are geared for professionals in the field, but they do have resources for the general public, available online at acs.org and agfd.sites.acs.org.

My employer is ARS, the primary research arm of the United States Department of Agriculture. ARS has nearly 100 facilities across the country, and one of the largest is the Eastern Regional Research Center in Wyndmoor, Pennsylvania, just outside of Philadelphia. The Center has six research units focused on aspects of agricultural research, including the Dairy and Functional Foods Research Unit, where I am. The Unit and its predecessors have conducted research on cheese and other dairy products since the late nineteenth century. Of our many advances over the years, the most visible to consumers is lactose-reduced milk, which will be discussed in Chapter 1. We also developed a reduced-fat Mozzarella that has been used in the National School Lunch Program since 1995. Nowadays, our work deals with cheese characterization and safety; whey utilization; conversion of milk casein to edible films; development of bacterial cultures for cheesemaking and bioactive peptides and proteins; lowering the environmental and economic impacts of food processing; and development of products from plant cell-wall polysaccharides in fruit and vegetable processing residues. ARS also has online resources for people who may not be scientists but who are interested in topics related to food (www.ars.usda.gov). Our agency is committed to public outreach, and has been for a long time. In fact, a book titled *Varieties of Cheese: Descriptions and Analyses* was first published by USDA in 1908, renamed *Cheese Varieties and Descriptions* in 1953 (this edition is available at the National Agricultural Library website), and revised again in 1978.

Many other books describe cheese varieties, along with recipes and how to buy and serve cheese. But this book will describe the chemistry, biology, physics, nutrition, and even climatology involved in cheese. It is arranged so that the more technical topics are confined to boxes that you can skip (I won't be insulted, honestly). Along the way, we'll encounter some related topics that will add to the discussion.

About one-third of the milk produced in the United States each year goes toward cheesemaking, and on average each American consumes around 34 pounds of cheese annually. Here is the story of the science behind it.

Acknowledgments

It's hard to focus the mind on praise, one thinks too much of the holes in the cheese, or the slice of cheese, of the emptiness that goes with all good.
—Beginning of letter from William Carlos Williams to fellow poet Marianne Moore, 1944

ACTUALLY, IT IS easy to praise the people who helped with this book. Thanks to Peggy Tomasula and Sevim Erhan for allowing me to write the book in the first place. Thanks to my other coworkers for their help over the years, especially Diane Van Hekken, Edyth Malin, Harry Farrell, and two long-time colleagues who unfortunately passed away several years ago, Virginia Holsinger and Phillip Smith. Edyth, Harry, and Ellie Brown are responsible for the molecular models. Ray Kwoczak is the cheesemaker at our lab (one of his make sheets is shown in Box 2.1), and Susan Iandola, Moushumi Paul, Jim Shieh, and Brien Sullivan are also on the project—thanks to all of them and to Peter Cooke and Doug Soroka, who produced the electron microscopy images. At the American Chemical Society, Samuel Toba and John Christensen invited me to give the webinar that started me on the path to this book. At Oxford University Press, Assistant Editor Hallie Stebbins invited me to write the book, and Associate Editor Mary Calvert and Editor Jeremy Lewis deserve my gratitude for getting it into publishable form, along with senior production editor Natalie F. Johnson, editorial assistant Erik Hane, and copyeditor Gail Cooper.

Thanks also to Sara Risch of Science by Design for advice on packaging, Linda Ettinger Lieberman at Moskowitz Jacobs, Inc., for checking the rules developing experimentation section of Chapter 14 and Hildegarde Heymann at University of California–Davis for information and opinions on complementary compounds. Thanks to Phil Wolff of the USDA's Dairy Grading Branch for checking the section on grading, to Nana Y. Farkye of California Polytechnic–San Luis Obispo for providing information on the World Championship Cheese Contest, and to Devin Peterson of the University of Minnesota for the information on gas chromatography-olfactometry and the photo. Special thanks to Paul McSweeney at University College Cork for reviewing the manuscript.

Thanks to Jim Victor and Mike Geno for allowing me to use pictures of their art work, and to Lois Tendler, Robert Sandell, and Karen Sandell-Stern for the milk bottle caps. Thanks also to my colleagues with the Division of Agricultural and Food Chemistry and my family for tolerating me. I have tried very hard to be as accurate as possible in this book; any errors are my fault. I also apologize if your favorite cheese isn't mentioned, but with thousands of varieties in the world, it is impossible to get to them all.

U.S. and Metric Units

As different as chalk and cheese.
—British expression referring to two things outwardly alike but not the same

U.S. CUSTOMARY UNITS (degrees Fahrenheit, inches, pounds) are used in this book since Americans are more familiar with those than with the metric system. The conversions to degrees Celsius, millimeters, grams, and milliliters follow.

Temperatures

The temperatures given in this book are in degrees Fahrenheit. The equivalents in degrees Celsius, rounded off to whole numbers, are given in Table 0.1.

Length

A micrometer is a millionth of a meter, or a thousandth of a millimeter. An inch is equivalent to 25.4 mm. Table 0.2 shows some conversions from inches to millimeters.

TABLE 0.1

Conversions between Fahrenheit and Celsius

°F	°C	°F	°C	°F	°C
32	0	76	24	120	49
34	1	78	26	122	50
36	2	80	27	124	51
38	3	82	28	126	52
40	4	84	29	128	53
42	6	86	30	130	54
44	7	88	31	132	56
46	8	90	32	134	57
48	9	92	33	136	58
50	10	94	34	138	59
52	11	96	36	140	60
54	12	98	37	145	63
56	13	100	38	150	66
58	14	102	39	155	68
60	16	104	40	160	71
62	17	106	41	165	74
64	18	108	42	170	77
66	19	110	43	175	79
68	20	112	44	180	82
70	21	114	46	185	85
72	22	116	47	190	88
74	23	118	48	195	91

TABLE 0.2

Conversions between inches and millimeters

in	mm	in	mm	in	mm
1/8	3	¼	6	3/8	10
½	13	¾	19	1	25

Mass, Concentration, and Volume

A gram contains 1000 milligrams, and a kilogram contains 1000 grams. A kilogram is 2.2 pounds, so a 1000-pound cow would weigh 455 kilograms.

A concentration of one part per million (ppm) is equivalent to one gram in 1000 kilograms, or a drop of water in 13 gallons. A part per billion (ppb) is one drop in 13,000 gallons—the volume of a round swimming pool 18 feet in diameter and 7 feet deep.

A gram of water is equivalent to a milliliter of volume (also called a cubic centimeter, or cc for short). Thirty milliliters make up a fluid ounce.

PERIODIC TABLE OF THE CHEESES

This table includes the categories of cheese, major cheesemaking countries, and some noted varieties mentioned in this book. Each box contains the atomic number and symbol for one of the 114 chemical elements whose existence has been confirmed.

1 **H** Havarti																	2 **He** Herve
3 **Li** Limburger	4 **Be** Beaufort											5 **B** Brie	6 **C** Camembert	7 **N** Norwegian Jarlsberg	8 **O** Oscypek	9 **F** French cheese	10 **Ne** Neufchatel
11 **Na** Netherlands cheese	12 **Mg** Manchego											13 **Al** Afuega'l Pitu	14 **Si** Stirred curd cheese	15 **P** Process cheese	16 **S** Swiss type cheese	17 **Cl** Cantal	18 **Ar** Asiago Pressado
19 **K** Kefalotyri	20 **Ca** Caerphilly	21 **Sc** Stretched curd cheese	22 **Ti** Tiroler Almkäse	23 **V** Very hard cheese	24 **Cr** Crottin de Chavignol	25 **Mn** Mahón-Menorca	26 **Fe** Feta	27 **Co** Cottage	28 **Ni** Natural rind cheese	29 **Cu** Casciotta d'Urbino	30 **Zn** Ziegenkäse	31 **Ga** German cheese	32 **Ge** Greek cheese	33 **As** Asadero	34 **Se** Selles-sur-Cher	35 **Br** Brick	36 **Kr** Kasseri
37 **Rb** Reblochon de Savoie	38 **Sr** Surface mold cheese	39 **Y** Yarg Cornish	40 **Zr** Zamorano	41 **Nb** Naboulsi	42 **Mo** Mozzarella	43 **Tc** Torta del Cesar	44 **Ru** Rocamadour	45 **Rh** Raschera	46 **Pd** Picodon	47 **Ag** Allgäuer Bergkäse	48 **Cd** Cheddar	49 **In** Interior mold cheese	50 **Sn** Spanish cheese	51 **Sb** Sbrinz	52 **Te** Tête de Moine	53 **I** Italian cheese	54 **Xe** Xigalo Siteias
55 **Cs** Cabrales	56 **Ba** Batzios	*	72 **Hf** Humboldt Fog	73 **Ta** Taleggio	74 **W** Whey cheese	75 **Re** Red Leicester	76 **Os** Ossau-Iraty	77 **Ir** Ibores	78 **Pt** Pont-l'Évèque	79 **Au** Austrian cheese	80 **Hg** Homogenized milk cheese	81 **Tl** Tilsit	82 **Pb** Pickled/ brined cheese	83 **Bi** Bitto	84 **Po** Portuguese cheese	85 **At** Anevato	86 **Rn** Romano
87 **Fr** Fresh cheese	88 **Ra** Ricotta	**	104 **Rf** Roquefort	105 **Db** Daniblu	106 **Sg** Spressa delle Guidicarie	107 **Bh** Berner Hobelkäse	108 **Hs** Hushåollsost	109 **Mt** Maytag Blue	110 **Ds** Dorset Blue	111 **Rg** Ragusano	112 **Cn** Chevrotin		114 **Fl** Formaella Arachovas Parnassou		116 **Lv** Livarot		

*	57 **La** Langres	58 **Ce** Cebriero	59 **Pr** Provolone	60 **Nd** Noord Hollandse Edammer	61 **Pm** Parmigiano Reggiano	62 **Sm** Smear-ripened cheese	63 **Eu** Époisses de Bourgogne	64 **Gd** Gouda	65 **Tb** Tiroler Bergkäse	66 **Dy** Derby	67 **Ho** Halloumi	68 **Er** Esrom	69 **Tm** Tomme de Savoie	70 **Yb** Yorkshire Blue	71 **Lu** Laguiole
**	89 **Ac** Abondance	90 **Th** Tronchón	91 **Pa** Picodon l'Ardèche	92 **U** U.S. cheese	93 **Np** Non-pasteurized milk cheese	94 **Pu** Pouligny-Saint-Pierre	95 **Am** Allgäuer Emmentaler	96 **Cm** Casu Marzu	97 **Bk** Berner Alpkäse	98 **Cf** Cheese food	99 **Es** English style cheese	100 **Fm** Fourme d'Ambert	101 **Md** Mont d'Or	102 **No** Noord Hollandse Gouda	103 **Lr** Ladotyri Mytilinis

THE SCIENCE OF CHEESE

Did you not pour me out like milk and curdle me like cheese?
—Job speaking to God, Job 10:10

1

IN THE BEGINNING

Milk

ONE WARM DAY in the Fertile Crescent some 8,000 years ago, a traveler started off on a journey with some milk in a pouch. The milk may have come from a sheep, which was one of the first animals to be domesticated, and the pouch was simply a sheep's stomach. After a while, the traveler stopped to take a drink from the pouch and discovered that the milk had separated into a soft white solid and a yellowish liquid. He took a taste of the solid and found it had a bland but agreeable flavor.

This is the most-told story about the first production of cheese; we know for certain that it was made in Northern Europe in the sixth millennium BC after archeologists analyzed residues in ceramic sieves found in modern-day Poland. Cheesemaking served as a way to preserve milk while providing nourishment to lactose-intolerant people (see below). Soon, others found that the flavor grew more intense the more days the cheese sat around; different species of animal produced different tasting cheese; and adding salt preserved it longer. The reasons behind all this are rooted in chemistry. The milk coagulated into a solid (the *curd*) and a liquid (the *whey*) because of enzymes in the stomach lining, and bacteria in the milk acted on the curd during storage to generate flavorful compounds. It took until the nineteenth century for anyone to realize this, but that didn't stop people from creating and experimenting with cheese. Now, more than 2,000 varieties of cheese are made around in the world. Some of today's varieties have been around for a while: Gorgonzola was first noted in AD 879, Roquefort in 1070, and Cheddar in 1500. More are being developed all the time, as cheesemakers apply trial and error and a combination of talent and luck.

The most important factors in cheesemaking are the composition of the milk, the development of acid as the curd forms, the amount of moisture in the curd, the way the curd is manipulated, and storage conditions. But before we continue, we should define some terms.

- **Moisture** is the amount of water in a food. Generally, the higher the moisture content is, the softer the cheese and the shorter the shelf life.
- **Fat** includes solid fats and liquid oils. A fat or oil molecule is in the form of a triglyceride, which consists of a "backbone" of glycerol bonded to three chains of *fatty acids*. If straightened out, the structure would resemble a capital E. The shorter fatty acid chains are volatile and contribute to aroma and flavor (one is depicted in Box 1.1 along with glycerol and a triglyceride). The fat in dairy products is often called *milkfat* or *butterfat*. Fats and oils are relatively stable in cheese, though they do degrade into flavor compounds. Cows produce less fat in summer than in winter. Feed also affects fat content in milk. Fat in milk is in the form of droplets, known as *globules*.
- **Protein** in cheese composes the network in which the fat globules and water are held. All proteins consist of *amino acids* linked together in a three-dimensional network, and typically almost all of the protein in cheese curd is *casein* (pronounced KAY-seen). Several other proteins are found in whey. Proteins will break apart with ripening, contributing to flavor and decreasing the hardness of the cheese. High heat or the addition of certain compounds will cause many proteins (though not casein) to unfold, or *denature*, resulting in a loss of biological activity. *Enzymes* are proteins that hook up with other molecules and speed up a chemical reaction. Milk has some 70 indigenous enzymes, and in most cases their role in cheese is unclear. We will see that non-indigenous enzymes added to the milk start the breakdown of casein and lactose.
- **Carbohydrate** in cheese is basically *lactose*, also known as *milk sugar*. Lactose will also break down to generate flavors. Milk is the only source of lactose found in nature.
- **Energy** refers to the caloric content of a food. A *food Calorie* (note the capital C) is the energy required to heat a kilogram of water by 1°C.
- **Minerals** include calcium and phosphorus in milk and the sodium chloride added to cheese.
- **Vitamins** are nutrients containing carbon, hydrogen, and oxygen that are required by the body and must be obtained from the diet. Vitamins A, D, E, and K are soluble in lipids, and the B vitamins and vitamin C are dissolved in water.

BOX 1.1
MOLECULAR STRUCTURES

Scientists find it easier to visualize molecules by drawing structures, using abbreviations for the chemical elements and lines for links between them. For instance, butanoic acid is a fatty acid found in cheese and rancid butter (the name comes from the Greek word for butter). Its structure fully written out looks like this:

$$\begin{array}{ccccccccccc} & & H & & H & & H & & & & \\ & & | & & | & & | & & & & \\ H & - & C & - & C & - & C & - & C & = & O \\ & & | & & | & & | & & | & & \\ & & H & & H & & H & & O & - & H \end{array}$$

Each C represents a carbon atom, each H a hydrogen atom, and each O an atom of oxygen. A line, either horizontal or vertical, is a single chemical bond, in which two adjacent atoms share an electron. Each carbon atom may be linked, or bonded, to as many as four other atoms, which in this book may be hydrogen, oxygen, nitrogen, sulfur, or another carbon. The double line to the right represents a double bond, in which adjacent atoms share two electrons. An O is capable of having only two bonds, which in food may be one double bond with carbon or two single bonds. An H is limited to one single bond, which it may share with carbon, oxygen, nitrogen, or sulfur. When a number of C's are linked together in a series of bonds, we'll refer to the string as a *carbon chain*.

When you get more complicated molecules, all those C's and H's clutter up the structure. We can streamline the picture by using numbers:

$$CH_3CH_2CH_2COOH$$

All we've done is remove the bonds, show the number of hydrogen atoms bonded to each carbon atom, and abbreviated the part with the oxygen atoms. Chemists know that COOH means that an O is double-bonded to a C, with an O-H also attached to the C. This COOH group tells us that the compound is an acid.

We can also depict the molecule without most of the H's, using a skeletal structure:

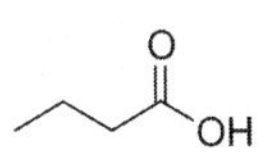

Here, we assume that the starting point on the left is a CH_3 and that every corner also corresponds to a C. The H's are only shown when their position is important or to avoid confusion. The H is shown with the last O here so we don't think something else is attached.

Glycerol, to which fatty acids are attached to form triglycerides, is a clear, syrupy substance often used in personal care products, where it is called "glycerin." Its skeletal structure looks like this:

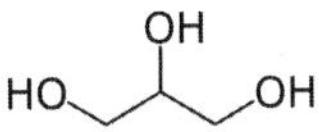

(*continued*)

BOX 1.1 (*continued*)

A triglyceride consisting of three butanoic acid molecules therefore can be depicted as:

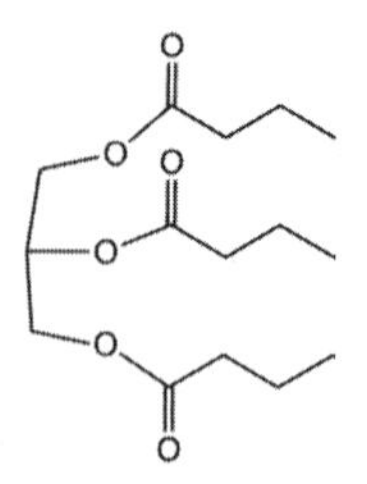

Molecules are not really shaped like any of these pictures. Instead, they bend themselves into the shapes requiring the least energy. Scientists use computer programs to find the most likely shapes, which they call *molecular models*. Some models of casein will be shown in Box 1.5.

Two more terms: for the purposes of this book, **artisanal** refers to cheese made primarily by hand in small batches from locally obtained milk using traditional methods. New varieties may be attempted, but the process is not fully mechanized. **Farmhouse** cheese is an artisanal cheese made on a farm from milk obtained from that farm's herd.

Mammals

Cows and other mammals have been used to make cheese since ancient times. Even the word *cheese* can be traced back a long way (Box 1.2). Almost 9.2 million milking cows are in the United States (Box 1.3), and six breeds of cow are responsible for 98 percent of the milk produced.

Table 1.1 lists the six cow breeds along with the amount and composition of the milk they produce. The last three columns are averages for Wisconsin cows.

Holsteins, also known as Holstein-Friesians, were developed in Holland and Germany more than 2,000 years ago. These are the familiar black and white animals that most people imagine when they think of cows. Holsteins are considered rugged and good feeders. They are the most efficient breed in generating large amounts of milk. Jerseys originated in the Channel Island of that name and are cream or light brown, sometimes with patches of white. They are sensitive and are finicky eaters, and are noted for producing more fat and protein than other breeds. Research in our laboratory showed that Jersey milk is more suitable for cheesemaking than the milk of the other five breeds; the more milk protein that is casein, the better the yield of cheese, and Jersey milk is the only one with over 83 percent of protein as casein. You can

BOX 1.2
ETYMOLOGY

Why is it good to be "the big cheese," but bad to be "cheesy"? Where does the word *cheese* come from? For these answers, we turn to etymology, the study of formation and development of words. *Cheese* derives from the Latin *cāseus*, passing through such forms as *chasī* (Old High German), *kasī* (Old Saxon), and *cēse* (Old English). Some other European languages have similar words for cheese: kaas (Dutch), käse (German), cáis (Irish), queso (Spanish), and queijo (Portuguese). *Casein* also comes from this word.

The Latin *jūs*, juice, evidently referring to the whey, led to northern European words for cheese: juusto (Finnish), juust (Estonian), and ost (Norwegian, Swedish, Danish, Icelandic).

A third Latin word, *formaticum*, means "shape" and was used in the term *formaticus caseus*, "shaped cheese." This gave us two other words for cheese: fromage (French) and formaggio (Italian).

The Greek word for cheese is *τυρί*, pronounced tyri. It is the source of the word *tyrosine*, an essential amino acid that was discovered in casein in 1846. It is also the root of *turophile*, "cheese lover."

Our English word *sour* comes from the Indo-European word *syr* or *sir*, relating to the souring of cheesemilk. The word is used for cheese in Slavic languages: сыр ("syr," Russian and Belarusian), сир ("seer," Ukrainian, Serbian, Croatian), сирене ("seerenneh," Bulgarian), ser (Polish), sýr (Czech), and syr (Slovak). Baltic languages also use syr/sir as the root: siers (Latvian) and sūris (Lithuanian).

The word for cheese in Hindi, Persian, and Turkish sounds like "paneer," which is also the name of a fresh South Asian cheese coagulated with a food acid such as lemon juice.

The expression *cheese it!*, meaning "stop it!" or "be quiet!" was featured in many old movies and comic books, often as "Cheese it, the cops!" It was noted in an 1811 dictionary of British slang and appears to have come from a mispronunciation of "cease." Another slang dictionary, from John Camden Hotten in 1913, cites "cheese your patter" and "cheese your barrikin," both of which basically meant "stop talking."

"Browned off," meaning disgusted or fed up, apparently originated with the Royal Air Force during World War II and referred to rusting of metal; the expression reminded some of browning of cheese when heated, leading to *cheesed off*.

How did *cheesy* turn into a word meaning cheap or inferior? Urdu, now a national language of Pakistan and an official language in five states of India, has a word *chīz* that means "thing." The British picked up on it when they controlled the area starting in the eighteenth century, and by 1818 incorporated it in a phrase "the real cheese," meaning "the real thing." *Hotten's Dictionary* defines cheese and cheesy as "a first-rate or very good article." In the U.S., a good fastball in baseball is sometimes called "hard cheese," with a high fastball being "high cheese" (Spanish players call it "alto queso"). Cheese also features in the expression "the big cheese," meaning a boss or very important person. But

(*continued*)

BOX 1.2 (*continued*)

sarcastic use of the word turned it into a pejorative, with the adjective "cheesy" being recorded in the negative sense in 1896. An unabridged 1940 dictionary has both senses for cheesy: "fine; excellent; smart; esp[ecially], ironically, worthless; good-for-nothing."

Many people's surnames are derived from the trade of their ancestors, and cheesemaking and dairying are no exceptions. The first individuals with the last names Cheeseman and Cheesewright were makers and sellers of cheese. The press used to squeeze the whey from the curds was called a wringer, giving us the last names Ringer and Wringer. Cowherd and Coward come from cow herder and Cowley from cow pasture (cow lea). Byers means cow sheds and Boothby is a farm with byers on it. One derivation of Day is dairy worker. "Wick" meant dairy farm in the thirteenth and fourteenth centuries, giving us Fenwick (dairy farm in the fen), Gatwick (goat farm), and Sedgwick (dairy farm in the reed beds). Cheese is still an English surname; the father of British comedian John Cleese was originally named Cheese, but changed it upon entering the army to avoid ridicule.

BOX 1.3
AMERICAN COWS

There were over six million farms in the United States in 1940, with three-quarters of them having at least one milking cow. Now America has around two million farms, and fewer than six percent of them have milking operations. The reasons for the decreases are:

- Technological advancement, starting with the introduction of electricity in rural areas, which enabled farmers to refrigerate their milk and keep larger quantities of it. Also, machinery has replaced workers, harvesting of crops and milking of animals has improved, the biology of plants and livestock is better understood, and computerized monitoring has given farmers a better idea of what to expect. Fewer farms are feeding more people.
- The shift from forage feeding to a confinement system with feed rations, which has increased milk production per cow and eliminated the time and effort required to shuttle cows between pasture and milking parlor. The amount of milk per cow was just 5,000 pounds a year in the 1940s.
- Specialization, as dairy farming is no longer a sideline on a farm along with several other activities. In the 1940s, about 45% of milk produced was for consumption on that farm, another 43% was sold as a sideline, and 12% was produced on and sold by a dairy-only farm. Now, about 75% of milk produced comes from dairy-only farms.
- The daily grind of having to stay on the farm and take care of cows is not appealing to people who desire easier ways to make a living.
- Economic considerations have forced many dairy farmers out of business. Almost 90% of dairy farms are family-owned, and some of them lack the money

BOX 1.3 (*continued*)

to withstand fluctuations in milk price. When milk prices are low and feed and fuel prices are high, some farmers are forced to sell their cows and equipment and close down. Others prosper by making farmhouse cheese, a value-added product.

The number of milking cows in the United States has dropped from 21,994,000 in 1950, to 10,799,000 in 1980, to 9,197,000 in 2011. The average number of cows per herd has grown from 6, to 32, to 179 in that span. Although the amount of milk produced by American cows continues to go up (it was over 196 billion pounds in 2011), the number of farms with a permit to sell milk has dwindled dramatically over the past twenty years, from more than 130,000 to less than 52,000.

Which state has the most dairy farms? That would be Wisconsin, with 12,100. Pennsylvania is a distant second, with 7,240. Table B1.3 shows the 11 states with at least 1,000 dairy farms as of 2011:

California averages a whopping 1,061 cows per herd, while the other states listed above average between 61 and 171. California, Wisconsin, and New York are the top three states in cow population, with Idaho (580,000 cows in 575 herds) fourth, Pennsylvania and Minnesota fifth and sixth, Texas (435,000 cows in 590 herds) seventh, Michigan eighth, New Mexico (333,000 cows in just 140 herds) ninth, and Ohio tenth. Farmers can have many cows in a herd if they have the land available, resulting in growth in nontraditional dairy areas with open space. Among the fifty states, Idaho ranked twentieth in milk production in 1975, and New Mexico was forty-first; both states have since become heavily involved in dairy farming.

TABLE B1.3

States with 1,000+ dairy farms

State	Dairy farms	Dairy cows
Wisconsin	12,100	1,265,000
Pennsylvania	7,240	538,000
New York	5,450	610,000
Minnesota	4,325	465,000
Ohio	3,170	269,000
Michigan	2,160	369,000
California	1,675	1,778,000
Iowa	1,670	201,000
Indiana	1,650	173,000
Missouri	1,530	94,000
Vermont	1,000	134,000

TABLE 1.1

Comparison of cow breeds

Breed	% of U.S. cows	Weight in pounds	Annual pounds of milk per cow	% fat in milk	% protein in milk
Holstein	89.0	1500	24,030	3.7	3.0
Jersey	7.6	1000	16,840	4.8	3.6
Brown Swiss	0.9	1400	18,800	4.1	3.4
Ayrshire	0.4	1150	15,700	3.9	3.2
Guernsey	0.3	1100	16,130	4.6	3.3
Milking shorthorn	0.2	1450	16,060	4.0	3.1

guess that Brown Swiss cows came from Switzerland and are dark brown in color. They are docile, though sometimes stubborn. Ayrshires, a red and white breed, were developed in the Scottish county of Ayr. Ayrshires are quite hardy but also nervous and can be difficult to manage. Guernseys, also named for the Channel Island where they were first bred, range in color from light fawn and white to deep yellow. Their milk is also yellow owing to its higher fat level and the secretion of orange-colored β-carotene from their feed (β, pronounced "beta," is the second letter of the Greek alphabet). Guernseys are efficient milk producers. Milking shorthorns originated in the English county of Durham and are known as dairy shorthorns in the British Isles and South Africa. They are a combination of red, white, and roan. They are considered to be efficient grazers and easy to manage.

Many other breeds are found outside the United States. You can figure out where Danish Red, Dutch Belted, Norwegian Red, and Russian Black Pied originated. Flamande and Montbéliard are French breeds, and Illawarra originated in the Illawarra region of New South Wales, Australia.

But you don't have to make cheese from cow's milk. Milk from many mammals can be used to make cheese, but the ability to collect the milk and coagulate it limits the choices to cows, goats, sheep, water buffalo, and a few others. As Cornell University Professor Frank V. Kosikowski, who was the dean of cheese science, once asked, "how does one milk a guinea pig or a 100-ton whale careering through heavy seas?"

Here (Table 1.2) are typical values for the composition of milk from dairy animals (there will be more comparisons in Chapter 11 and an explanation of scientific names in Chapter 9):

TABLE 1.2

Comparison of milk from different mammals

Species	Scientific name	Water	Fat	Protein	Lactose
			% IN MILK		
Cow	*Bos taurus*	87.4	3.6	3.2	4.7
Goat	*Capra hircus*	87.3	3.8	3.4	4.1
Sheep	*Ovis aries*	80.1	7.9	6.2	4.9
Water buffalo	*Bubalus bubalus*	82.8	7.4	3.8	4.8
Reindeer	*Rangifer tarandus*	66.7	18.0	10.1	2.8
Yak	*Bos grunniens*	82.7	6.5	5.8	4.6
Zebu	*Bos indicus*	86.5	4.7	3.2	4.7

American dairy goat breeds include Alpine, LaMancha, Nubian, Saanen, and Toggenburg. The two major dairy sheep breeds in the United States are East Friesian and Lacaune. Many water buffalo are in Italy, where the milk is used for authentic Mozzarella cheese, or *mozzarella di bufala*. A couple of herds of river buffalo (as opposed to swamp buffalo, the other species of this animal) are maintained in the United States for the same purpose.

Reindeer (known as caribou in Alaska and Canada) have a great deal of protein and fat in their milk—baby reindeer need extra nourishment and fat insulation in their cold environment. Reindeer are the only milk source for people living in Arctic regions since no other dairy animal can survive up there. Two people are required to milk this animal, one of whom has to hold it by the horns. Finland's Juustoleipä ("bread cheese") often comes from reindeer, and is caramelized before an open fire after pressing.

Yak and zebu are in the same genus as cows. Yak cheese originated in Nepal, and is also made in Bhutan, Kashmir, Tibet, and Mongolia. Zebus originated in South Asia, have spread to other parts of that continent, and have been imported to Africa, Brazil, and Mexico.

There is one farm, in Sweden, making moose cheese. It takes two hours to obtain a gallon of milk from each of their three animals. They produce about 650 pounds of cheese a year, and it sells for $500 a pound, making it the world's most expensive cheese.

As you might imagine, the composition of milk varies among species. Box 1.4 compares cow, goat, sheep, and water buffalo milk. They do have one thing in common: they all produce milk in the same way.

When a dairy animal consumes food, it is digested in her stomach and small intestine. The stomach consists of four chambers: the rumen, reticulum, omasum, and abomasum. Semi-digested food (the cud) is regurgitated from the rumen and re-chewed. The reticulum is only partially separated from the rumen, allowing the contents of both to mix. The food then passes into the omasum, where some of the volatile fatty acids that were generated from carbohydrates in the rumen are absorbed into the bloodstream. The abomasum is the only part of the stomach containing gastric juices, and prepares dietary protein for further digestion and absorption in the small intestine.

BOX 1.4
MILK COMPOSITION

The Agricultural Research Service maintains an online database of food composition, and here is a chart based on those numbers for milk along with other sources. Amounts are per 100 grams (3.5 ounces) of milk. Units are in milligrams, except where noted in parentheses. NM means not measured.

Vitamin C is susceptible to degradation by heat, and up to a quarter of it is lost by pasteurization. The level of D too low to prevent deficiency of this vitamin, so American processors add it to milk at a concentration of 400 international units (10 micrograms) per quart. Vitamin D is created in the skin upon exposure to sunlight, but that process is insufficient for our needs in the darker winter months. It is present in fatty fish, liver, and other organ meats, but these foods are not consumed in great quantities in the United States. Some yogurts, margarines, and breakfast cereals are also fortified with vitamin D.

TABLE B1.4

Milk composition of four species

Component	Cow	Goat	Sheep	Water buffalo
Water (grams)	88.13	87.03	80.70	83.39
Protein (grams)	3.15	3.56	5.98	3.75

BOX 1.4 (*continued*)

Component	Cow	Goat	Sheep	Water buffalo
Carbohydrate (grams)	4.80	4.45	5.36	5.18
Fat (grams)	3.25	4.14	7.00	6.89
Cholesterol	10	11	27	19
Energy (Calories)	61	69	108	97
MINERALS				
Calcium	113	134	193	169
Copper	0.025	0.046	0.046	0.046
Iron	0.03	0.05	0.10	0.12
Magnesium	10	14	18	0.31
Phosphorus	84	111	158	117
Potassium	132	204	137	178
Sodium	43	50	44	52
Zinc	0.37	0.30	0.54	0.22
VITAMINS				
Retinol, A (international units)	162	198	147	178
Thiamine, B_1	0.046	0.048	0.065	0.052
Riboflavin, B_2	0.169	0.138	0.355	0.135
Niacin, B_3	0.089	0.277	0.417	0.091
Pantothenic acid	0.373	0.310	0.407	0.192
Pyridoxine, B_6	0.036	0.046	0.060	0.023
Folate	0.005	0.001	0.007	0.006
Cobalamin, B_{12} (micrograms)	0.045	0.07	0.71	0.36
Ascorbic acid, C	0.94	1.29	4.16	NM
Calciferol, D (micrograms)	0.06	0.058	0.18	NM
Tocopherol, E	0.07	0.07	NM	NM
Phylloquinone, K (micrograms)	0.3	0.3	NM	NM

Nutrients absorbed into the bloodstream are carried to the mammary gland, which contains alveoli. An alveolus is a cavity or sac with cells on its surface. Each alveolus has many secretory cells in which proteins, lactose, and fat are built. Amino acids from the digested proteins are assembled into casein and whey proteins, and the caseins become balled together into casein micelles (MY-sells, see below). At the same time, glucose is converted to lactose, and triglycerides are constructed from fatty acids. Some cells may specialize in the different processes.

In each alveolus, the secretory cells surround a lumen, a reservoir in which components mix to form milk. From here on, the milk is mostly water with casein micelles and globules of fat suspended in it. Each lumen drains milk into a duct, and each duct leads to a cistern inside a teat. When the animal is stimulated by its sucking offspring or by simply being in the milking parlor, she releases the hormone oxytocin. The hormone signals the muscles to compress the alveoli, and the milk stored in the lumen is forced into the ducts, through the teat cistern, and out.

A dairy cow typically gives birth to her first calf at two years of age. The gestation period of a cow is nine months, and the lactation period, the time when a cow produces milk, averages 305 days. A calf needs only 2,200 pounds of milk, so in earlier times the rest was taken by the farmer. These days, calves are fed a diet and the farmer recovers all of the milk. Milk production reaches a maximum 40–60 days after calving, at which point the cow is bred again. The second calf is born a year after the first one and the cycle repeats. Cows are capable of having ten lactations, but are usually culled from the herd after two or three.

Casein Micelles and Fat Globules

Casein micelles are up to half a micrometer in diameter, and many (but not all) dairy scientists theorize that each micelle is made up of units known as submicelles, which are held together by *colloidal calcium phosphate*. A *colloid* is a particle between a tenth and a thousandth of a micrometer in size, and suspended in a liquid. *Phosphate* is an atom of phosphorus surrounded by four oxygen atoms; several amino acids in casein contain phosphate groups. Phosphate bonds with calcium in milk to form calcium phosphate, which cements the submicelles together. The formula for colloidal calcium phosphate in milk is believed to be $Ca_3(PO_4)_2$, so we'll just abbreviate it as CCP. Newly born animals (and people) need calcium and phosphorus to grow, and casein micelles allow plenty of it to be available in the form of CCP. When cheese is made, the acid generated by the bacteria releases some of the CCP to the whey; the amounts of acid produced and CCP lost are determining factors in the type of cheese being made.

The four major types of casein are α_{s1}-casein (pronounced alpha-ess-one), β-casein, κ-casein (kappa), and α_{s2}-casein. Intact κ-casein is the only one of the four not sensitive to calcium, and is found on the outside of the casein micelle where it prevents the other casein micelles from clumping together (Box 1.5). We will see in Chapter 2 that κ-casein molecules are cut into two pieces during the early stages of most cheesemaking, causing the casein micelles to fall apart and coagulate into a curd. The inability to coagulate casein is the major reason why milk from some mammals cannot be made into cheese (Box 1.6).

Unlike casein micelles, fat globules do aggregate in untreated cow's milk. Before homogenization caught on, milk was sold in bottles with agglomerated fat globules—cream—forming a layer on top in half an hour. The cream rises because

BOX 1.5
CASEIN MODELS

Casein micelles contain the four major types of casein, with four α_{s1}-casein and four β-casein molecules for every κ-casein and α_{s2}-casein. Molecular modeling has shown that each type of casein has a particular structure. The molecules appear to fit together within a submicelle, though not everyone is convinced of that idea; I once attended a conference where someone asked, "Do you believe in submicelles?" and got a mixed response.

The shapes of α_{s1}-, β-, and κ-casein are shown on the left, and, if you are a believer, a model of a submicelle with these molecules is at top right. Scanning electron micrographs showing magnified casein micelles and submicelles are at lower right.

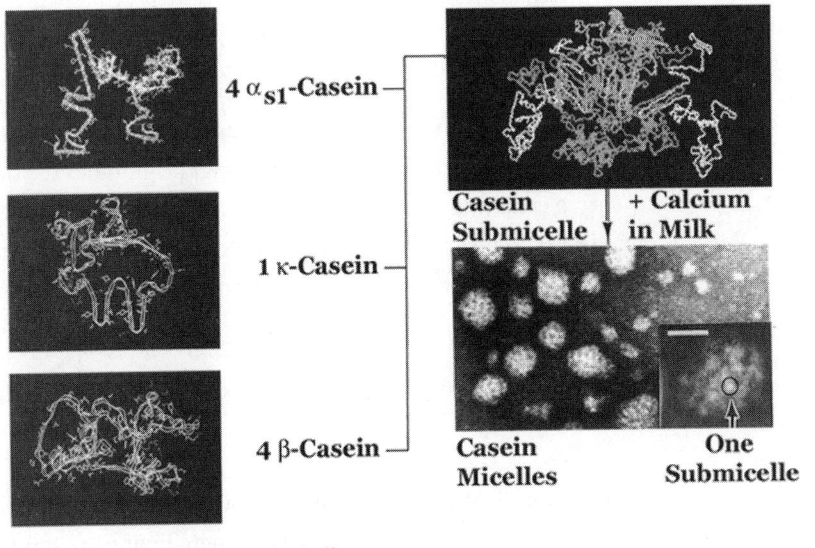

Shapes of casein molecules and micelles.

BOX 1.6
MILK FROM OTHER SPECIES

Although all mammals produce milk, cheese cannot be made easily, or at all, from some species. Camels have historically supplied milk to North Africans and Asians, but it is difficult to coagulate the milk into a curd, and much of the protein and fat are lost in the whey, so cheesemaking is not economical. Central Asians ferment horse milk into the alcoholic drink *kumis*, but the milk lacks κ-casein and will not coagulate at all. Some mammals, such as zebras, cannot be domesticated, while others, such as rabbits, are too small to produce much milk.

Can you make cheese from human breast milk? That is the strangest question I have ever been asked about cheese, and it's been asked several times. The answer is no, because the milk won't coagulate properly. Breast-feeding mothers often have milk left over, and people sometimes wonder what can be done with it. Perhaps make some sort of cheese from it? Daniel Angerer, chef of Klee Brasserie in New York City, prepared and served some in his restaurant in 2010 before the Health Department asked him to please stop it. It turns out that his recipe called for 2 cups of breast milk and 2 cups of conventional milk per batch. Breast milk alone can't be made into cheese because it contains too little protein (only 1%), and only some of the protein is casein. In early lactation, 90% of the proteins are whey proteins, decreasing to 50% in late lactation. Research in our laboratory some years ago demonstrated that human milk will not form a curd.

A French web site, Le Petit Singly, advertises that they have made cheese from "breast milk of woman" since 1947. If you try to order some online, you will be sent to a page asking you if you are crazy. They admit on that page that it is a hoax, though some news outlets have reported on the "company" without discovering it is a joke.

If you do have surplus breast milk, it is best to donate it to a breast milk bank, where it will be used to feed infants in neonatal intensive care units.

it is less dense than the rest of the liquid. Homogenization reduces the size of the fat globules and strips off their protective membrane, causing the globules to stick to the casein and preventing creaming. On the other hand, milk from goats, sheep, and water buffalo takes hours to cream. As a result, some goat and sheep farmers refer to their animals' milk as being "naturally homogenized." Most cheese is made without homogenizing the milk, as we will see in Box 2.2.

Other Uses for Milk

You can do other things with milk aside from drinking it or making it into cheese, yogurt, etc. The whey is often powdered and used as a food ingredient or as a nutritional supplement, as will be discussed in Box 2.9. If the casein is separated by adding rennet and then hardened in a formaldehyde solution, it is converted to a plastic that was formerly used for buttons, imitation ivory, and fancy goods. The casein may also be separated by adding acid to skim milk. If you mix this "acid casein" with sodium bicarbonate (baking soda), water, denaturing agents, and a few other ingredients, you get sticky, flowing white glue.

Centuries ago, before commercial paint was available, farmers would coat the bare wood of their barns with a mixture of milk, lime, and linseed oil. The milk-lime combination resulted in a binding agent similar to the casein–baking soda blend, and the oil acted as sealant that allowed the paint to soak into the wood. To prevent fungi from growing on the barn, they added a natural fungicide—rust. Rust, or iron oxide, turned the paint red, which is why barns are often that color. In more recent times, some dairy farmers began painting their barns white, knowing that consumers associate milk with that color.

Lactose Intolerance and Milk Allergy

Some people have problems digesting milk. Milk allergy is an adverse reaction to one or more proteins in milk, whereas lactose intolerance is the inability to break down the lactose molecule. These conditions are definitely not the same: allergy, or hypersensitivity, is mediated by the immune system, and intolerance is not. Cow's milk proteins are the initial foreign proteins fed to many infants, and may provoke an allergic reaction in the immature digestive system. Around 2.5 percent of children have milk allergy, but most of them outgrow it by age three. It is rare in adults. To determine the presence of milk allergy (or any dietary allergy), the suspected food is eliminated from the diet. If symptoms disappear, the food may be given in gradually increasing amounts until symptoms reappear, thus giving a range in which the food may be consumed without adverse effects.

Lactose intolerance affects far more people—roughly 65 percent of the world's population. The others are termed *lactase persistent*, because they retain the lactase enzyme (β-galactosidase) that splits the lactose molecule into glucose and galactose (Box 1.7). Lactose cannot be absorbed by the body unless this split occurs in the small intestine. If lactase is not present, the lactose moves to the large intestine where bacteria metabolize it, generating uncomfortable gas in the form of carbon dioxide, hydrogen, and methane. The lactose also draws water from the intestinal wall, causing diarrhea.

The ability to digest lactose appears to involve evolutionary and demographic factors along with genetic, physiological, and social aspects. Lactase persistence apparently arose as a result of a mutation of a particular gene some 7,000 to 10,000 years ago within dairy farmers in Central Europe. Vitamin D is necessary for absorption of calcium and phosphate, and milk serves as a good source for both in the higher

BOX 1.7
LACTOSE METABOLISM

The lactose molecule consists of two rings of carbon, oxygen, and hydrogen atoms joined by an oxygen atom. The enzyme β-galactosidase (pronounced BAIT-ta ga-lac-toe-SY-dase) breaks the molecule at the oxygen junction, forming galactose and glucose, both of which can be digested by humans. In this diagram, the various parts of the molecules are pointed in diverse directions, and many of the hydrogen atoms are not shown. Can you spot the difference between glucose and galactose?

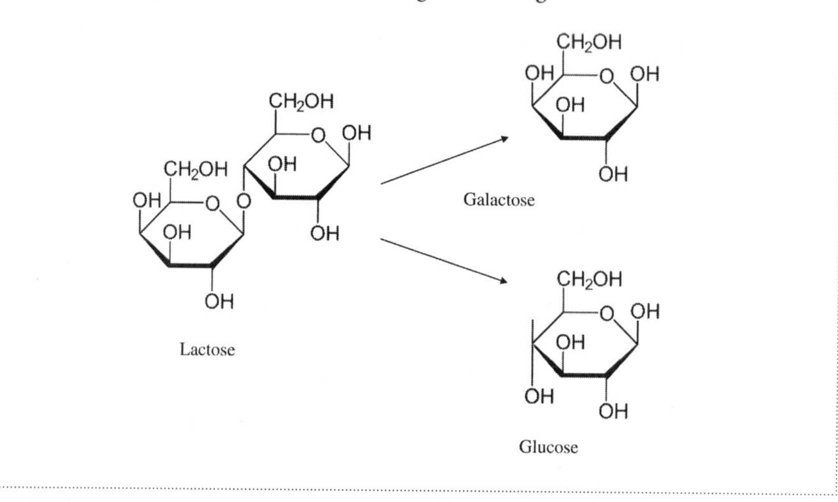

latitudes of Eurasia where the production of vitamin D in the skin is hampered by lower levels of sunlight. The following table (Table 1.3) shows the distribution of lactase persistence in various countries.

You can see that almost everyone in northwestern Europe is lactase persistent, but the number decreases as one goes south and east. The genome sequence of "Ötzi the Iceman," who died 5,300 years ago and was frozen in an Italian mountain pass until he was found in 1991, indicates that that he was lactose intolerant. Milk fat residues in pots found in Libya indicate processing of cow's milk there between 5200 and 3800 BC, supporting a theory that lactase-persistent people and their cattle moved from Europe into Africa during that period. The migrants presumably intermarried with the locals, causing at least some of their offspring to acquire the lactase-persistence gene, which is useful in hot climates where milk serves as a relatively safe source of water. As a result, some African and Middle Eastern populations can digest milk. Most Eastern and Southern Africans cannot, and many in Asia and Australia are also unable to do so.

TABLE 1.3

Lactase persistence around the world

Country	Population	Lactase persistence (%)
	EUROPE	
United Kingdom	British	95–97
Ireland	Irish	96
Denmark	Danes	96
Finland	Finns	83–92
Germany	Germans above 52°N latitude	87–94
Germany	Germans below 52°N latitude	76–86
Spain	Galicians	66
Poland	Poles	59–71
Italy	Northern Italians	49
Italy	Sardinians	11–18
Russia	3 European groups	41–52
	MIDDLE EAST	
Turkey	Turks	26–31
Jordan	Urban/agricultural Arabs	76
Jordan	Mediterranean Arabs	23
Saudi Arabia	Bedouins and urban residents	81–86
Saudi Arabia	Arabs	43
Kuwait	Kuwaitis	42–53
	AFRICA	
Niger	Tuareg	87
Senegal	4 groups	71–100
Sudan	Haddendoa, Amarar	80–87
Sudan	Dinka, Nuba	21–25
South Africa	Sotho and Swazi	25–35
South Africa	Xhosa and Zulu	11–19
Uganda, Zambia	Bantu	0–6
	ASIA AND PACIFIC	
Pakistan	Punjabi	41
India	Indians	36–39

(*continued*)

TABLE 1.3 (*continued*)

Country	Population	Lactase persistence (%)
Sri Lanka	Sri Lankan	28–29
China	Kazakh	24
China	Mongols and Northern Han	8–12
Taiwan	Chinese	12
Thailand	Thai	0–3
Australia	Aborigines	16
New Zealand	Maori	36

Fortunately, people can do two things to avoid lactose problems when consuming dairy products. The first is to drink lactose-reduced milk. This product was developed in our laboratory in the 1970s and the principle is simply the inclusion of β-galactosidase in the milk. The enzyme is added to milk after being generated by yeast such as *Kluyveromyces lactis* and molds such as *Aspergillus oryzae* and *Aspergillus niger*. Lactose-reduced and lactose-free milk are available from several manufacturers.

The other way to avoid lactose-related problems is to eat cheese. As we will see, nearly all of the lactose in cheese is removed during manufacturing and aging. The next chapters describe how this happens.

Heavens defend me from that Welsh fairy, lest he transform me to a piece of cheese!
—WILLIAM SHAKESPEARE, *Merry Wives of Windsor*, 1602

2

CURDS AND WHEY

Cheesemaking

NOWADAYS, CHEESE IS produced by men and women instead of fairies. The procedure for making cheese depends on the variety, and the cheesemaker keeps track using a "make sheet" (Box 2.1). Here are the basic steps:

- Standardize milk
- Pasteurize milk
- Add starter culture
- Add rennet
- Cut and cook curd
- Drain whey
- Pile curds
- Handle curds
- Add salt
- Press
- Coat or Package
- Ripen

An important factor in cheesemaking is the *yield*, the amount of cheese obtained from milk. Cheddar manufacture typically results in 10 pounds of cheese per 100 pounds of milk, for a 10 percent yield. Other semi-hard and hard

BOX 2.1

MAKE SHEET

Cheesemakers use make sheets to keep track of the steps, timing, and other information during a cheese manufacturing run. An example for Queso Fresco is below. T/A in the next-to-last column refers to titratable acidity, a measure of the lactic acid generated.

QF041410 C

Cheese Make on April 14, 2010 — **Queso Fresco - Phase III**

4 kg curd for Lm study, rest to be stored at 4 or 15C for 60 d - measure quality traits

Received unpastrized milk on 04/13/10 — Actual raw	milk fat:	6/7 °C
Fat: 35/35	pH: 6.8	0.16 T/A
Store overnight at 4°C.		
Morning of cheese making	milk fat:	
Standardized to: 3.5% fat	3.5/3.5	0.16 T/A
Past. milk at 63 °C for 15 sec. hold — After past.	milk fat:	
cool and transfer to cheese vat	3.5/3.5	0.16 T/A
Charm test results:	Pass/Pass	

Actual time	TIME hr:min	OPERATION	TEMP °C	pH:	T/A	Room temp
12:10	00:00	Load Kussel vat with milk 182.6 Kg	6°C	6.7	0.149	9.5
12:40	0:30	Milk reaches 32°C	32.0°	6.54	0.150	20.9°C
12:40	:30	Add $CaCl_2$ 182.6 g $CaCl_2$ in 500 ml DI H_2O, final 0.1% wt milk; mix at least 5 min	32.0°	6.28	0.149	
12:45	:35	Add rennet 12.7 ml rennet in 40 ml DI water; mix thoroughly, then allow to coagulate	32.1°			
1:05	:	Cut curd using 3/8" knives	32.0°	6.21	0.12	
1:10		Heal for 5 min	32.2°			
1:10	:	Raise temp to 39°C, 1°C/5 min	32.1°	6.18	0.119	
1:20	:	1:15 33° / 1:20 34° Check after 10 min	34.2°			21.5°C
1:30	:	1:25 35° / 1:30 36° Check after 20 min	36.3°	6.15	0.108	
:	:	1:35 37° / 1:40 38° Check after 30 min	38.2°			
1:40	:	Cook for 80 min at 39°C 1:45	39.0°	6.18	0.099	
1:55	:	Check after 15 min	38.9°	6.13	0.099	
2:10	:	Check after 25 min	39.1°	6.13	0.099	
2:15	:	Drain whey - remove about 90 kg whey	39.5°	whey: 6.15	whey: 0.099	Vol whey:
2:25 2:30 2:35	:	Add 2.63 Kg salt (12.5% salt per gal milk) add salt in 3 stages with 5 min of mixing after each application; then drain completely		5.93 curd:	total →	95.05 Kg 56.30 Kg 151.35
2:40	:	Trench curds, drain for 15 min.		curd: 5.95	whey: 0.098	
2:55	:	Check after 15 min		curd: 5.96	whey: 0.098	Room temp
3:00		Cut thin (3 inch high), place on trays, cover with cheese cloth, refrigerate until core is < 70°F	total curd	28.14 Kg		
3:35		Divide into 5 portions curd wt before:	70°			PM
:	:	a) 0 no milling	5.5 Kg			
		b) mill using plastic plate - no holes	5.44 Kg			
		c) mill using smaller plate - 8 holes	5.40 Kg			
		d) mill using old grinder - 5 holes x2	10.68 Kg			
5:00		total curd	26.92 Kg			20.0
:	:	Tightly pack rest of curd into molds.				Room temp
6:00	:	Flip mold over, place in 4°C refrig overnight		curd: 26.91 Kg		Room temp
Next day	AM	Remove from mold, cut into large manageable blocks				AM
10:00 AM		fa, fb, fc blocks each cut into 1/3 for 4°C (wk 1, 4, 8)				
11:00 AM	:	double batch of d mill, cut into 6 portions - half to 4 and 10°C (wk 1, 4, 8)				

A make sheet for Queso Fresco.

cheeses vary from 8 to 12 percent yield; cottage cheese, which contains 80 percent water, produces a 15–17 percent yield. The cheesemaker will adjust procedures and amounts to maximize yield (and profit margin) while maintaining the best possible product.

Collection and Standardization

When milk is collected at the farm, it has to be immediately cooled (see Box 2.2 for the grades of milk and Chapter 16 for rules and regulations). This raw milk is transferred to an insulated tanker truck (it does not have to provide refrigeration) and delivered to the cheese plant unpasteurized. The fat and protein levels of the milk may then have to be adjusted to create a uniform product with the highest possible yield, in a process is known as *standardization*. Skim milk, skim milk powder, or

BOX 2.2

MILK STANDARDS

All dairy farmers have to abide by quality standards for animal health, facility cleanliness, milking procedure, equipment cleaning procedures, water quality, and waste disposal. No drug residues or added water are allowed. The milk is inspected regularly and judged to be either Grade A (fluid grade) or Grade B (manufacturing grade). The liquid milk purchased and consumed in the United States is subjected to relatively little processing, so its standards therefore have to be very high. About 97% of American cow's milk is Grade A, which meets these highest standards. Grade B milk is not consumed until after it has been converted into cheese, butter, or milk powder; the processing involved in making these products eliminates any dangers from lower-quality milk. Grade B milk cannot be used for fluid milk, but Grade A can be used for making cheese and the other products.

Milk normally contains relatively few bacteria as it leaves the cow, but the number will grow quickly without refrigeration. The storage temperature and resulting bacterial count are the primary difference between the two grades. Grade A milk is cooled to a maximum of 50° within 4 hours of the start of milking, and to a maximum of 45° within 2 hours of the completion of milking. Less than 100,000 bacteria per milliliter (3 million per ounce) of Grade A milk is allowed. Grade B milk is cooled to a maximum of 50° unless it is taken to the processing plant within 2 hours; if stored in a tank, it is cooled to 40° within 2 hours and maintained at 50° or less. The maximum number of bacteria in Grade B milk is 500,000 per milliliter.

(*continued*)

BOX 2.2 *(continued)*

Grade B milk bottle cap from the 1930s.
Source: Author's collection.

BOX 2.3
HOMOGENIZATION

The fat droplets in raw milk are various sizes (1–15 micrometers), and may be considered as dispersed throughout skim milk. Milk fat is less dense than water and therefore rises to the top. In homogenization, milk is forced through a small orifice at high pressure (2000–4000 psi, or 13.8–27.6 megapascals), reducing the size of the fat droplets to 1–2 micrometers, making them more uniform. The milk is sent through twice to prevent the droplets from clumping. By doing this, the milk fat globule membrane that surrounds each droplet is stripped away. Proteins in the milk then cover the droplets, making them denser and less likely to rise to the top of the milk container. Homogenization started to become popular in the 1930s and was almost universal by the 1960s; prior to that, milk was delivered to homes with a cream layer on top.

Cheesemilk is not usually homogenized because the reduction in size of the fat globules alters the texture and functionality of the resulting cheese. Smaller globules are trapped in the protein matrix and are not released during heating, which prevents melting. Removal of the milk fat globule membrane leaves the globules susceptible to a higher level of degradation, which may lead to rancid flavors. Blue,

BOX 2.3 (*continued*)

Feta, and cream cheese are often made from homogenized cow's milk since it leads to a whiter color, increased breakdown of fat (desirable in these varieties), and a smoother texture. Other varieties come out with a poor texture and a moisture content that is undesirably high when the milk is homogenized. Queso Fresco is supposed to be a non-melting cheese, so the milk is homogenized to prevent melting.

cream may be added to the milk, or some fat may be removed. The milk is not usually homogenized (Box 2.3).

Pasteurization

Once the milk is standardized, it is pasteurized by heating to 161° for 15 seconds, or 145° for 30 minutes. These time–temperature combinations (some others with shorter times and higher temperatures are not normally used for cheese) must be applied to "every particle of milk" according to the regulations in the United States, and will inactivate all milk-borne pathogens and most of the naturally occurring

FIGURE 2.1 Pasteurization heat exchanger.
Source: Author's collection.

BOX 2.4
RAW MILK CHEESE

For many years, authorities such as the Centers for Disease Control (CDC) and the U.S. Food and Drug Administration (FDA) have feared that raw milk can be dangerous because there is no pasteurization step for killing pathogens that might be present. Illnesses caused by *E. coli* O157:H7, *Campylobacter*, and *Salmonella* have been linked to consumption of raw milk and products such as raw milk cheese. Listeriosis, an uncommon but often fatal illness caused by *Listeria monocytogenes* may also be correlated with raw milk and cheese made from it. Interstate sales of raw milk have been forbidden in the United States since 1982, and 20 states will not allow sales at all. Retail store sales are currently legal in 12 states, and on-farm sales are allowed in another 13 states. The remaining states have other regulations such as "cow shares," which entail part-ownership of a cow and the right to any raw milk coming from her. Much raw milk comes from certified dairies that have to undergo stringent sanitation requirements and testing.

In the U.S., cheese made from raw milk must be held for at least 60 days. Any pathogens in the milk should die out during this time, and the acidity generated by the cheesemaking process will prevent pathogens from the environment from taking hold. Bacterial contamination from just one cow can spread through an entire vat of milk and then to a production run of raw milk cheese, which is why the rules exist. Raw milk cheeses from outside the U.S. must also follow this requirement or they cannot be imported. The CDC and FDA recommend that children under five, pregnant women, and people with immunodeficiency problems or chronic disease should not consume raw milk products.

Many cheese connoisseurs disagree with these regulations, since pasteurization not only kills off pathogens but also the naturally occurring bacteria that contribute to flavor and might convey health benefits. They often refer to pasteurized milk cheeses as "dead cheese." A test of 890 attendees at a recent Oregon State University food festival showed that 426 people preferred the raw milk version of a cheese, 319 preferred the pasteurized version, and the rest had no preference or could not tell the difference. The study also showed that some consumers consider raw milk cheese to be a delicacy and are willing to pay more for it.

enzymes. Pasteurization is performed with a heat exchanger, such as the one shown in Figure 2.1. Raw milk is often used for cheesemaking outside the United States (and sometimes in it; see Box 2.4). But unless you specifically look for raw milk cheese, the product will be pasteurized.

When the milk is ready, it is pumped into a temperature-controlled stainless steel vat. The vats are often elevated in large operations so the curds and whey can drain out the bottom by gravity (Figure 2.2). The vats are double-walled, meaning that water at the desired temperature flows between the outer and inner walls.

FIGURE 2.2 Elevated double-walled cheese vats. The disk-shaped objects atop the vats cover the stirring motors. Heating/cooling water flows from the vertical pipes into the space between the walls.
Source: Author's collection.

Additives

The cheesemaker may put a few legal additives into the milk. One is *annatto*, which comes from the *achiote* tree found in Mexico, the Caribbean, and the western portion of South America. Annatto gives cheese an orange or more pronounced yellow color (Box 2.5). Conversely, titanium dioxide may be added to mask color. It is a mineral commonly added to toothpaste—and to Mozzarella, leading to a whiter cheese. Hydrogen peroxide and benzoyl peroxide are added to some varieties to bleach color. Calcium chloride ($CaCl_2$) may be added to speed up coagulation and improve the firmness and yield of the cheese. Fruits, vegetables, meats, spices, flavors, and smoke may also be added during cheesemaking, as described in Chapter 14.

Classes of Cheese

At this point, we should take a look at the different categories of cheese, since each is made a different way. Cheeses may be classified according to moisture content, country of origin, species of animal, texture, or other ways, but we will use manufacturing procedure: see Table 2.1.

BOX 2.5

WHY IS CHEESE YELLOW?

> *"They had but one last remaining night together, so they embraced each other as tightly as that two-flavor entwined string cheese that is orange and yellowish-white, the orange probably being a bland Cheddar and the white . . . Mozzarella, although it could possibly be Provolone or just plain American, as it really doesn't taste distinctly dissimilar from the orange, yet they would have you believe it does by coloring it differently."*
>
> —MARIANN SIMMS, 2003 winner of the Bulwer-Lytton Fiction Contest for writing intentionally bad opening sentences to imaginary novels

The color of food influences our perception of it. Studies have shown that color affects our expectations of the flavor of food, more than labeling and even its actual taste. When a transparent version of a popular cola beverage was introduced in 1992–1993, regular drinkers of the normally colored soda did not like it, even though the flavors were identical. Consumers anticipated the flavor from the soda's appearance, and were thrown off when they drank it. The product was discontinued.

Cheese is no exception to the association between color and flavor. Some people perceive orange Cheddar to be richer than yellow Cheddar, though their fat contents and flavors are identical, so cheesemakers add annatto to color it. The major component of annatto is bixin (the scientific name of the achiote tree, the source of annatto, is *Bixa orellana*). Much of the molecular structure of bixin is similar to that of β-carotene, the compound that gives carrots their orange color. Cows transfer carotenoids (β-carotene and related compounds) from their diet to the milk, where they bind to the fat. The fat content of milk is less than 4%, the carotenoid concentration is less than 0.1%, and the fat globules are surrounded by casein, so the yellow color is not visible. Most of the whey is lost during cheesemaking, causing the fat and carotenoid contents to increase, and the casein network loosens up to reveal the fat, to the point where the resulting cheese takes on a yellow color. Goats, sheep, and water buffalo do not pass carotenoids to milk (converting it to vitamin A instead), and their cheese is white. Consumers expect some cow's milk cheeses to be white or light yellow, resulting in the addition of titanium dioxide, a widely-used white pigment.

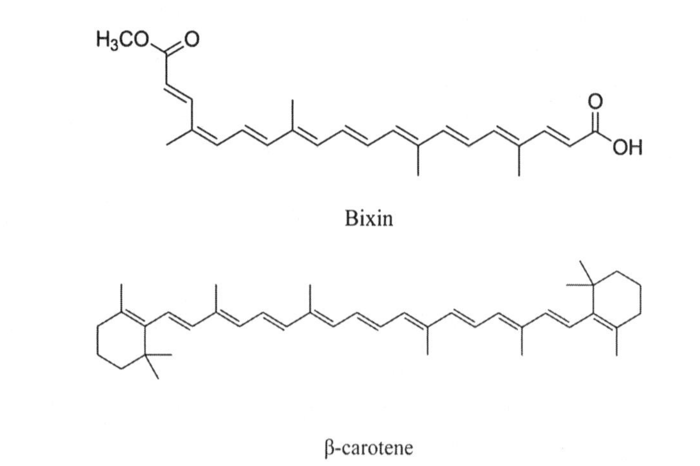

Bixin

β-carotene

TABLE 2.1

Classes of cheese

Class (alternate name)	Varieties commonly sold in the U.S.
Fresh soft	Cottage, cream
Whey	Ricotta
Pickled (brined)	Feta
Stretched curd (pasta filata)	Mozzarella, Provolone
Surface mold (bloomy rind)	Brie, Camembert
Smear-ripened (washed rind)	Limburger, Brick
Interior mold (blue-veined)	Blue, Stilton
Cheddared (English style)	Cheddar, Cheshire
Washed curd	Gouda, Colby
With eyes (Swiss type)	Emmental, Gruyère
Very hard	Parmesan, Romano

As you might imagine, differences in factors such as microorganisms, temperature, and timing result in the great diversity in cheese throughout the world. Let's start with the starter.

Starter Cultures

When the milk has cooled to the desired temperature, the acidification process is begun. For fresh soft cheeses such as cottage and cream cheese, a food-grade acid such as citric acid, vinegar, or lactic acid may be added to coagulate the casein. Raw milk contains naturally-occurring bacteria, so unpasteurized cheeses are sometimes acidified by allowing the bacteria to grow without human intervention. Otherwise, starter culture is added, and the milk is stirred to mix it. Starter cultures consist of lactic acid bacteria, which ferment the sugar in milk—they convert lactose to lactic acid and make the milk more acidic (Box 2.6). These bacteria

- allow the milk to coagulate faster when rennet is added, since rennet enzymes work better under acidic conditions,
- cause the protein matrix in the cheese to contract and force out more whey,
- make it difficult for non-lactic acid bacteria to survive, and
- contribute greatly to flavor and texture formation during ripening.

BOX 2.6
LACTIC ACID

Lactic acid is formed from glucose and galactose in a long series of reactions involving more than a dozen enzymes in lactic acid bacteria. Each lactose molecule generates four lactic acid molecules, so the acidity of milk containing starter culture can build up quickly. Acidity and alkalinity are measured by pH level, with 7 being neutral, lower numbers being acidic, and higher numbers being basic or alkaline. The pH of milk is around 6.65, meaning it is slightly acidic. The structure of lactic acid is:

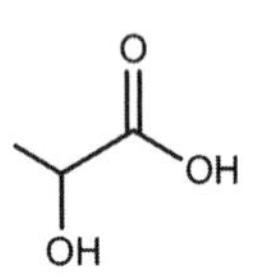

In spite of what has been popularly believed for almost a century, some experts now theorize that your muscles do not generate lactic acid when you overwork them. Acid builds up when your muscles become fatigued, but this may be caused by other cellular processes. Lactic acid is, however, produced by bacteria in your mouth, and is responsible for tooth decay.

One other acid found in milk is citric acid, which is found in citrus fruit and also in milk. It has this structure:

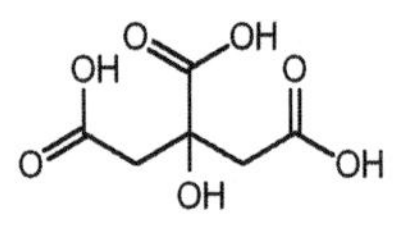

Both lactic and citric acids are precursors of some flavor compounds, as we will see in Chapter 4.

Starters are available as powders, liquids, or frozen liquids, and are added to the vat as "direct vat starters." Many cheese plants grow their own "bulk starters," starter cultures that are propagated in reconstituted skim milk powder or other media. Cultures may be "defined strain," containing as many as six known and well-defined strains of bacteria, or "mixed-strain," with an undefined mix of bacteria. Since bacteria survive in whey, some places use the previous day's whey in a procedure charmingly called "backslopping."

Starters fall into two groups. *Mesophilic* (Greek for "middle-loving") bacteria tolerate moderate temperatures (68°–104°) with an optimum of 86°. *Thermophilic* ("heat-loving") bacteria thrive at higher temperatures (86°–122°, and optimally around 108°). Common mesophiles include various subspecies of *Lactococcus lactis* along with *Leuconostoc mesenteroides* and other Lactococcus species (*Lactococcus* and *Leuconostoc* are genus names, and *lactis* and *mesenteroides* are species names. We will discuss the difference between genus, species, subspecies, and strain in Chapter 9). Thermophiles are used in Italian and Swiss varieties where the initial

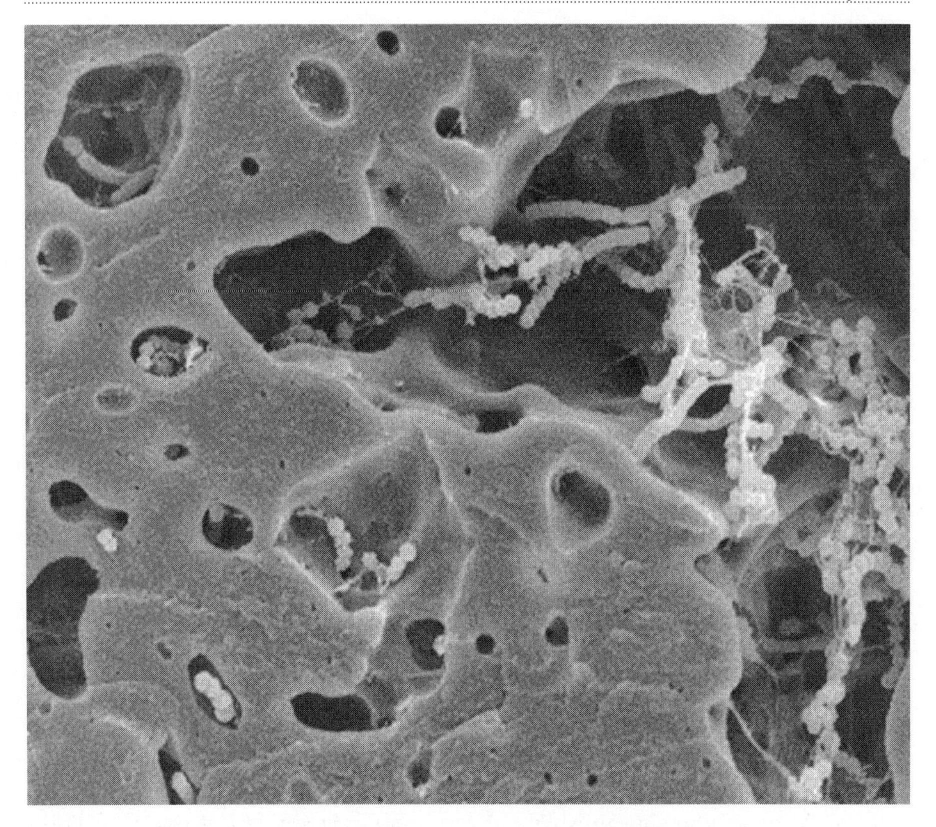

FIGURE 2.3 Starter cultures in Mozzarella cheese. The round bacteria, like strings of pearls, are *Streptococcus thermophiles*, and the rod-shaped bacteria, like sausages, are *Lactobacillus delbrueckii*, subspecies *bulgaricus*. The gray area is the casein matrix, and the dark cavities had been occupied by fat globules.
Source: Author's collection.

temperatures are high, and include *Lactobacillus delbrueckii* subspecies *bulgaricus, Lactobacillus helveticus*, and *Streptococcus thermophilus* (Figure 2.3). Leuconostoc and *Lactococcus lactis* also metabolize citrate, which is present in milk at low concentrations. The breakdown products of citrate are important contributors to flavor, and the carbon dioxide (CO_2) gas generated during the process helps with formation of the eyes in some cheeses such as Edam and Gouda.

Adjunct (or secondary) cultures are also added when making some varieties. These are not responsible for producing acid but instead generate flavors during ripening. They are often added after the curd has formed. The three types of adjuncts are:

- Bacteria such as Propionibacteria and Coryneform bacteria. *Propionibacterium freudenreichii* metabolize citrate, lactate, and other compounds, and produce carbon dioxide gas that leads to the "holes" in Swiss, Gruyère, and other varieties. Coryneforms, primarily *Brevibacterium linens*, are found on the surface of cheeses such as Brick and Limburger.

- Molds such as *Penicillum camemberti* and *Penicillum roqueforti*, unsurprisingly used for Camembert and Roquefort. *P. roqueforti* is also used to make blue cheese, Stilton, and others.
- Yeasts such as *Geotrichum candidum* and various Candida species, used in Brie, Limburger, and some cheeses with mold.

Lipase enzymes are also added in the preparation of some cheese varieties such as Parmesan. Lipase attacks the fat in milk and leads to flavor formation.

But cheesemaking is not as easy as pouring in a preparation of microorganisms and waiting for perfect curd to be created. Cheesemakers have to be microbe wranglers who are sensitive to changes in the performance of their tiny little friends. A corollary of Murphy's Law states "under the most rigorously controlled conditions of pressure, temperature, volume, humidity, and other variables, the organism will do as it damn well pleases." The most dreaded condition is *bacteriophage*, or simply *phage* (can rhyme with either "cage" or "lodge"). Bacterial cells can be infected with viruses just like human cells, and phage results when this happens to a starter culture. The milk acidifies normally during a phage infection, but the culture stops growing, and the cheese produced has to be discarded. Phage-resistant cultures are available, and the cheesemaker usually rotates between two separate culture stocks.

In case you are a bit squeamish about eating bacteria, remember that your body has ten times as many bacterial cells as human ones. Over 500 species are found in your intestines alone. You literally can't live without them. Cheese provides an environment for microbes to grow, resulting in a product that runs a gamut of flavors and textures. But not all of the bacteria in the curd are placed there deliberately. Box 2.7 describes non-starter bacteria in cheese and Box 2.8 describes an unexpected side-effect that may occur with them.

BOX 2.7

NSLAB

This acronym stands for *non-starter lactic acid bacteria*. These are the bacteria that are not inactivated by pasteurization, or otherwise find their way into the milk and curd. Some NSLAB are resistant to the cleaning and disinfecting agents used in the cheese plant, and have been isolated from the equipment, floors, and drains. NSLAB may also be airborne, originating from the environment around the plant. NSLAB proliferate during ripening, eventually becoming dominant in the cheese. At the start of ripening, a typical cheese may contain a billion colony-forming units of starter bacteria per milliliter of cheese. During the first month, that number may decrease to ten million, allowing NSLAB to increase. They can provide typical flavors, atypical but desirable

BOX 2.7 (*continued*)

flavors, and off-flavors. Some 60 *Lactobacillus* species are known, of which about a dozen have been detected in cheese as NSLAB.

The presence of NSLAB, molds, and other microorganisms that are not intentionally added brings some uncertainty to the cheesemaking process. Consider the story of Liederkranz cheese, which is an American version of the European cheese Limburger. Liederkranz was developed in Monroe, New York, in 1891 by a Swiss cheesemaker named Emil Frey, who went on to create Velveeta 27 years later. Frey named the cheese after the Liederkranz ("wreath of song") singing society. The Monroe Cheese Company, the only firm to manufacture this variety, moved to Van Wert, Ohio, in 1926. Frey, his assistants, the starter cultures, and the original equipment all moved, too. The wooden shelves used to store the cheese stayed put, however. When they tried to make the cheese at the new place, it did not turn out properly. Various changes were attempted, but the original could not be duplicated. Finally, they smeared the new tile walls with the residue from the wooden shelves from the old plant. Nobody would ever do that sort of thing nowadays, but the cheese came out perfectly. Some of the microorganisms (probably Coryneform bacteria) floated from the wall and into the vats, the same way they wafted from the shelves and into the vats at Monroe. It was then that the cheesemakers realized that their product was not just affected by what they add, but also by things they don't add.

BOX 2.8

GREEN CHEESE AND GREEN MEAT

> *"The moone is made of a greene cheese."*
> —THOMAS HEYWOOD, *Proverbs*, 1546

When you say that the moon is made of green cheese, you are not claiming that Earth's satellite is green in color. You are actually comparing the color and shape of the moon to a fresh or unripened cheese: a green cheese is an unaged one that is white to light yellow. The expression originated in England long before Heywood recorded it.

But cheese can turn processed meat green. When packaging some cheeses with cured meat inside a sandwich, the meat begins to turn green within a few hours or days. The culprits are starter bacteria that produce H_2O_2, hydrogen peroxide. H_2O_2 resembles H_2O, water, but the extra oxygen atom makes it highly reactive.

Meat is cured by adding sodium nitrite, $NaNO_2$, either in pure form or by adding celery juice or celery powder, which are naturally high in it. The $NaNO_2$ reacts with myoglobin, the protein that gives muscle its red color, and when heated forms nitrosohemochrome, the pink pigment in cured meat. H_2O_2 reacts with this compound, creating green-colored products. A commonly used starter culture, *Leuconostoc mesenteroides*, is a

(*continued*)

BOX 2.8 (*continued*)

very effective producer of H_2O_2, and another popular starter, *Lactobacillus delbrueckii* subspecies *bulgaricus*, is also a good H_2O_2 generator. At least one nonstarter lactic acid bacteria (*Weissella paramesenteroides*, found in grasses and silage) also forms H_2O_2. These bacteria tolerate NaCl, $NaNO_2$, heat and smoke, and the H_2O_2 they generate remains even if the bacteria are killed during processing. Changing the cheese used in the sandwich will eliminate the green meat problem.

Rennet

The *pH* of a material is the measure of its acidity or alkalinity, on a scale of 0 (highly acidic) to 14 (highly alkaline), with 7.0 being neutral. Milk has a pH around 6.6, meaning it is slightly acidic. Once the pH of the milk has decreased by 0.2–0.3 units, rennet is diluted in water and added to coagulate, or set, the milk into curd. The initial curd is called the *coagulum*. Rennet consists of a number of enzymes, especially *chymosin* (also called *rennin*), which is produced in the stomachs of mammals for digestion of the mother's milk. Chymosin is a *protease*, or proteolytic enzyme, meaning that it breaks down protein and forms a curd. Other enzymes in rennet are *pepsin*, which is also a protease, and *lipase*. Rennet is typically added when the milk is at 104°–107°, though some varieties have higher set temperatures. The amount of rennet added is usually about 26 ounces per 1,000 gallons of milk.

Traditionally, rennet was isolated from calf stomachs, such as the ones shown in Figure 2.4. Some cheese is still made this way, especially outside the United States. Professor David Barbano of Cornell University has visited a cheese plant in Italy where they leave an entire stomach at the bottom of the vat. When the enzyme is used up, a new stomach replaces the old one. (No, this practice is never seen in the United States.)

Most cheese plants use other rennet preparations. Some rennets are secreted by certain fungi such as *Cryphonectria parasitica* and *Rhizomucor miehei*. Others use recombinant chymosin derived from microorganisms, including *Kluyveromyces lactis* or safe versions of *E. coli*. Some European cheesemakers use rennet derived from local plants, such as the cardoon (or artichoke thistle). The cardoon is used to coagulate Portugal's principal cheese, Serra da Estrela, which is made from raw sheep's milk. Cardoon also coagulates cow's milk, but is not used because the resulting cheese is bitter. Fungal, recombinant, and plant rennets have the advantage of not being from animal sources, making them acceptable to people with dietary restrictions.

FIGURE 2.4 Calf stomachs drying in preparation for cheesemaking.
Source: FAO.

BOX 2.9
COAGULATION

One milliliter of milk contains hundreds of trillions of casein micelles, which consist of submicelles bound together by CCP. The surface of these micelles is composed of submicelles rich in κ-casein, which is a string of 169 amino acids. One portion of the κ-casein molecule is attracted to water and sticks out of the micelle surface like a hair. The hairs of the κ-caseins seem to prevent micelles from interacting with each other.

Two of the amino acids in κ-casein, phenylalanine and methionine, are found at positions 105 and 106. Chymosin breaks the bond between them, sending the macropeptide (amino acids 106–169) and the protruding hairs into the whey. (The 1–105 portion is *para-κ*-casein, which remains in the curd). After most of the macropeptide has been removed from the κ-casein, the micelles are able to stick together when they collide, stay in contact, and build into a coagulum.

Coagulation is highly dependent on temperature and must take place above 65°. Casein aggregation also depends on the pH level: all of the CCP goes from the micelles to the whey at pH 4.6, but only half of it does at pH 5.6. Additional factors may include size and shape of the molecules, electrostatic charges on the micelles, and other forces. The mechanism of rennet coagulation is not completely known and has drawn interest from many researchers, since it is such an important part of making cheese.

FIGURE 2.5 The author using cheese knives to cut curd.
Source: Author's collection.

All rennets work in the same manner, by attacking casein. As mentioned in the previous chapter, casein comes in the form of micelles, made up of submicelles and stabilized by κ-casein. Chymosin snaps the κ-casein at a bond between two particular amino acids; the larger section of κ-casein (known as *para*-κ-casein) stays attached to the submicelle, and the smaller part (macropeptide) is dissolved in the whey. The micelle is thus destabilized, and various interactions then allow the caseins to aggregate into a three-dimensional coagulum (Box 2.9). Chymosin is water-soluble, so the more moisture in the curd the more chymosin is retained in the cheese, leading to extra flavor development later on.

Cutting and Cooking Curd

Once the curd has formed, the cheesemaker begins the process of removing whey from it. With some softer cheeses such as Camembert, the curd is gently ladled into boxes and allowed to drain by gravity overnight. Harder cheeses require cutting of the curd with *cheese knives*, consisting of a metal frame strung with horizontal wires and another frame with vertical ones, as shown in Figure 2.5. Quarter-inch knives, for instance, have wires that are a quarter-inch apart. By sweeping through the vat horizontally and vertically, the curd is cut into cubes, increasing the surface area

FIGURE 2.6 Stirring the curds and whey (bottom).
Source: Author's collection.

(slicing a piece of something exposes new surfaces) and allowing more whey to be released. Expulsion of whey from curd is known as *syneresis* (sin-er-EE-sis). The curd pieces are gently stirred to keep them circulating without shattering them (Figure 2.6), and the vat is heated, *cooking* the curd and causing the casein to fuse into filaments. The variety of cheese being made dictates the curd size, temperature, and heating time.

The cook temperature is important because it:

- Regulates the starter culture. Bacteria are active within a certain temperature range.
- Helps control removal of calcium. The more active the starter culture bacteria, the more acidification takes place, and the more soluble the CCP, which is dissolved in the whey. Cheese with a low pH and low calcium usually has a more crumbly texture. It also affects rennet activity. High cook temperatures inactivate rennet and prevent the enzymes from breaking down protein during ripening.
- Controls whey removal from the curd. More whey is expelled at higher temperatures.
- Minimizes growth of pathogenic and spoilage microorganisms in the vat. Higher temperatures are more likely to accomplish this.

The temperature inside the vat is slowly raised to the cook temperature; if raised too quickly, the outside of the curd pieces will dehydrate and harden, preventing whey from escaping from the inside.

Whey Drainage

Eventually the pH of the whey drops to 6.1–6.4. Cheeses such as Swiss are initially pressed before the whey is drained, which leads to a smooth texture without openings due to whey pockets or trapped air. Another type of opening is starting to be created, however, since the Propionibacteria in the starter culture are generating CO_2 gas from lactate digestion. Gas bubbles form in the curd, the smaller bubbles merge to make larger ones, and the characteristic holes, known as *eyes*, are formed. With other varieties, the whey is drained out from the bottom of the vat, leaving the curds exposed to air. Box 2.10 describes what happens to the whey.

Handling Curds

By now the casein filaments are fusing, or matting, into a solid mass. Some cheeses without eyes are subjected to *cheddaring*, in which slabs of curd are cut, stacked, and flipped to give time for the pH to drop while creating pressure and allowing heat to be retained and more whey to be removed (Figure 2.7). Cheddar, Queso Fresco, and

BOX 2.10
WHEY UTILIZATION

The whey from renneted cheese curd, called sweet whey, used to be discarded, fed to pigs, or spread on fields as a fertilizer. In 1942, one company poured their whey down a deep abandoned well on their property, apparently deciding that where there's a well, there's a whey. The trapped whey, under pressure, eventually developed enough gas to blow off the top of the well.

Whey is now too valuable to toss out. By allowing it to acidify and heating it nearly to boiling, the whey proteins coagulate into a fine curd that is made into Ricotta or Requesón. Boiling whey with milk and cream causes the lactose to caramelize, producing the Norwegian Geitost. Many Mozzarella plants also have a Ricotta operation. But the real money-making use for whey is in the manufacture of whey powders. After the fat is removed, whey is filtered under pressure through microfiltration membranes that catch bacteria and casein, and the whey proteins are retained by an ultrafiltration membrane. After drying to remove the water, whey protein concentrate (20%–89% protein) or whey protein isolate (at least 90% protein) is obtained. Whey powder is used in many foods because it foams well, provides a golden surface on baked goods, binds water, and adds nutritive value. Our laboratory has used whey protein powder to produce high-protein cheese puffs.

BOX 2.10 (*continued*)

In Chapter 1, we mentioned that proteins come in the form of amino acids linked together. Three amino acids, valine (shown below), leucine, and isoleucine, have branches or forks. These branched-chain amino acids are found in great numbers in muscle tissue, and fuel working muscles while stimulating the generation of protein. Whey proteins are about two-thirds β-lactoglobulin, one-fourth α-lactalbumin, and one-twelfth serum albumin, and all of these contain 19%–25% valine, leucine, and isoleucine. These proteins are rapidly absorbed into the body. Nutrition and health food stores sell canisters of whey protein powder to body-builders and athletes. People lose muscle mass as they become older, and whey protein is helpful for reducing frailty.

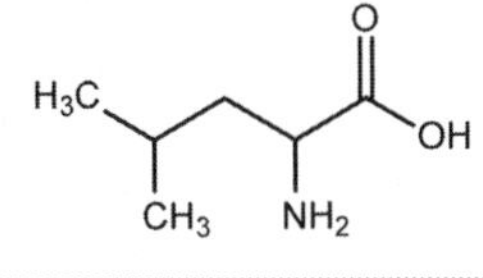

FIGURE 2.7 Curds piled for whey drainage.
Source: Author's collection.

FIGURE 2.8 Milling the curd.
Source: Author's collection.

other varieties are *milled*—sending the curd through a machine that chops the curd into pieces (Figure 2.8).

Salting

Salt, more properly *sodium chloride* or *NaCl*, is added to cheese to enhance flavor and to control microbial and enzymatic activity. The acid production of the starter cultures and growth of spoilage bacteria are inhibited by NaCl, which causes dehydration of bacterial cells. With some varieties, curds are *dry-salted* in the vat by sprinkling a measured amount of NaCl over them. Blue cheeses are dry-salted by rubbing NaCl on the surface, and Mozzarella, Feta, and others are immersed in brine containing approximately 22 percent NaCl, at a pH of about 5.1 and a temperature around 54°. The NaCl content of cheese varies, but is usually under 2.5 percent. There will be more on salt in Chapter 5.

Pressing

The curd is transferred into rectangular boxes or round hoops for pressing. These stainless steel forms are also known as *molds*, but we won't use that term, to avoid confusion with the microorganism. The forms are lined with cheesecloth,

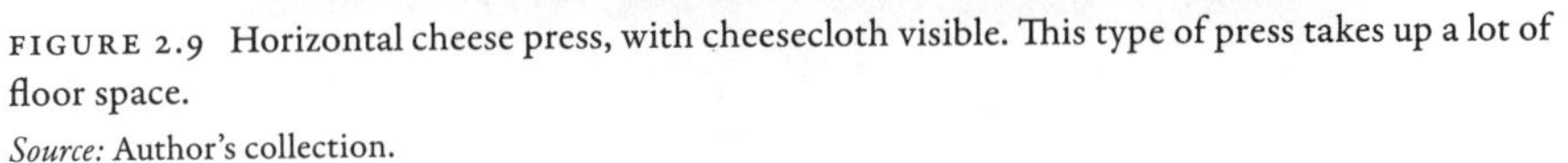

FIGURE 2.9 Horizontal cheese press, with cheesecloth visible. This type of press takes up a lot of floor space.
Source: Author's collection.

a cotton gauze invented for this purpose, and pressure squeezes the last of the free whey through the cloth and out through small holes in the forms (Figures 2.9 and 2.10). This step also shapes the cheese and helps impart the correct texture. Soft cheese is barely pressed at all, and Cheddar is pressed at 25,000 pounds per square inch (172 megapascals, in the metric system). A cheese with a disk shape is known as a *wheel*, and a cheese with rectangular sides is called a *block* or *loaf*.

Cheese does not have to be pressed and formed to be enjoyed. Fresh salted curds are a snack food in some places (see Box 2.11).

Coating or Packaging

Cheese is packaged to reduce moisture loss from the surface and prevent contamination. Softer varieties are packaged in a wooden container to allow for stacking. Aluminum foil, cellophane, paper, plastic wrap, and even wet towels are often used on cheese. Since CO_2 and other gases may be generated in some cheeses, the wrap should be permeable to prevent bulging of the package in

FIGURE 2.10 Vertical cheese press in use. This type of press has a small footprint but creates more pressure at the bottom than at the top.
Source: Author's collection.

these varieties. According to Dr. Sara Risch, a packaging expert who runs Science by Design in Chicago, vacuum packaging is the most effective in keeping air out, and curtails further microbial activity in the cheese (Figure 2.11). Some cheeses are aged uncovered so that the surface forms a rind. The surface mold and smear-ripened cheeses have a rind containing microorganisms, and hard cheeses such as Stilton and Parmesan obtain a rind through drying out of the surface. Rinds help control release of gas from the cheese and absorption of air. Many other cheeses are protected from the environment to prevent a rind from forming. Some cheeses, especially those made from goat's milk, are coated with ash to prevent the formation of a rind.

The ash may also assist with growth of desirable molds that increase the pH of the cheese by metabolizing amino acids, and lower the acidic flavor.

BOX 2.11

WHY DO FRESH CHEESE CURDS SQUEAK?

Some small cheese companies, especially in the northern U.S., sell bite-size cheese curds that have not been pressed or aged. These curds will emit a squeak when you bite into them, especially if they are less than a day old. The casein network in unpressed fresh curd has not knitted into a compact structure, but rather is porous with plenty of air trapped inside. The wet, elastic curd vibrates when you put your teeth through it, and this vibration is at a frequency in the audible range.

Cheese curds.
Source: Author's collection.

You can observe a similar effect by pulling your wet hair between your fingers after shampooing it. The detergent in the shampoo removes the oil from your hair, so your fingers don't glide smoothly, but "skip" along wet strands. The outer surface of hair is also protein, primarily keratin, but this time it is your fingers that do the vibrating.

School students used to hear a high-pitched squeal when the teacher ran a piece of chalk across the blackboard in a certain manner. (If you are under 25, ask your parents what a blackboard is). When the chalk was positioned at a particular angle and gripped the right way, it would skip rapidly. The resulting high-frequency noise would reverberate in the ear canal and produce a physical reaction in many students.

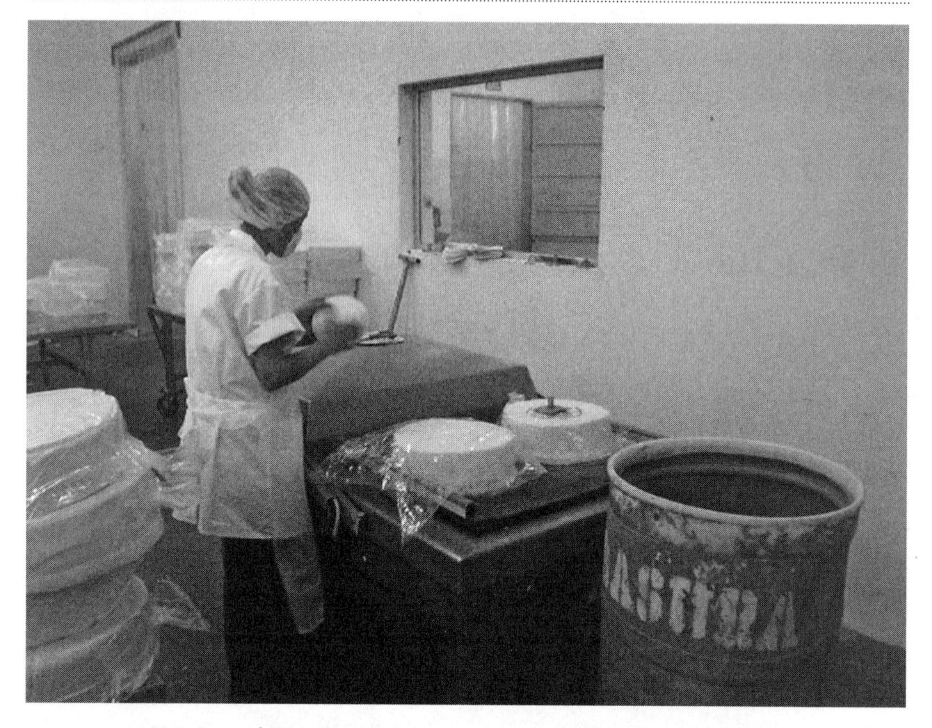

FIGURE 2.11 Vacuum packaging.
Source: Author's collection.

Shredding

Natamycin, which inhibits mold and fungus growth in the moist conditions inside the pouch, is also added. Natamycin is produced by a common soil bacterium and has been used for years on sausages, packaged salads, and other foods. Cheddar, Mozzarella, and other cheeses that may be used for melting are often shredded and sealed in plastic pouches. The shreds are prone to clumping, so they are coated with a small amount of cellulose, corn starch, potato starch, or calcium sulfate. These powders coat the shreds and block out the moisture that causes caking.

Aging

Once the cheese has been made, it has to be stored for a length of time in a step called *aging* or *ripening*. Different varieties necessitate different temperatures, relative humidity levels, and storage times, which affect growth of microorganisms and thus the rate and extent of flavor and texture development. Unpackaged cheeses need specific conditions to control formation of a rind or development of yeast or mold on the surface. Many varieties require two or even three rooms to properly age the cheese. This topic will be covered in Chapter 3.

After the cheese is brought home, it is best to keep it between 42° and 52°, in darkness, with the humidity around 80 percent. Softer cheeses require a refrigerator, which is typically 37°–41°. If a cheese has been opened but not fully eaten, wrap it with parchment or waxed paper and then enclose it all with plastic or foil. Cheeses that come in a box or tub should be stored in it. If the cheese is odiferous or runny, it may be wrapped in paper and placed on a plate, and the plate then covered with plastic. Wrapping should be changed every day or two, since repeated opening of the refrigerator results in temperature fluctuations that are detrimental to cheese. Freezing cheese has been described as an assault on a living thing and is not advisable—the texture is damaged when the water turns into ice crystals and tears into the protein matrix.

Shape

The shape of a cheese variety has its origins in the traditional methods for manufacturing it. Dry-salted cheeses require a good deal of pressure to shape the curd into a solid and non-fragile shape. Hoops were less likely to deform or break than rectangular forms, and were easier to manufacture. Swiss-type cheeses were made with adjustable hoops that could be tightened or loosened. In recent years, stronger stainless steel forms began to be used, since the resulting cheeses require less space in storage. At the retail shop, the wheels are sectioned into wedges like a pie.

Soft cheeses, which need less pressing, are often rectangular. The corners can be a problem with rectangular cheeses. Surface mold cheeses are round to prevent the microorganisms from attacking corners in four directions (top, bottom, and two sides). A disk-shaped cheese has only a top, a bottom, and one side going all the way around. Sharp corners tend to be over-salted (with NaCl penetration from four directions), resulting in differing characteristics than the bulk of the cheese block. Some cheeses are therefore pressed in square hoops with rounded corners. The thickness of a cheese is also important for absorption of salt. The larger the dimensions of a cheese, the longer it will take for salt to penetrate through it evenly.

Many traditional cheese shapes are used. Provolone has four: truncated cone, melon, pear, and sausage. Brick is unsurprisingly shaped like a brick, and Swiss types come in wheels. A cheese that has a rectangular cross-section is frequently referred to as a "loaf." British and Australian cheesemakers fashion some cheeses into cloth-bound barrel shapes known as "truckles." French goat's milk cheeses are shaped into disks, drums, fig shapes, logs, and narrow pyramids. Sainte-Maure de Touraine is fashioned into logs with a straw through the center to hold it together and allow air inside. Ireland's St. Killian is similar to Camembert, but is hexagonal. Colby and Cheddar sometimes come in cylinders, known as "Longhorns." Scamorza is often fashioned into shapes of animals.

Edam produced for export is coated with an impermeable red paraffin wax that makes the cheese resemble a red cannonball. The red color originally came from the application of turnsole dye, which was used as a food colorant and to illuminate medieval manuscripts. Rags soaked with the dye were hung over pans of urine, and the ammonia gas that emanated therefrom made the dye alkaline, turning it purple. The cheeses were rubbed with the rags, and the surface became red upon drying. Eventually red wax was substituted, to retain the traditional color. The spherical shape reduces surface area, allowing more moisture to be retained, and eliminates corners that could be chipped off during handling.

BOX 2.12
GIANT CHEESES

"He has bitten his way into the Big Cheese."
—O. HENRY, *The Unprofitable Servant*, 1911

Giant cheeses have been manufactured for publicity or special occasions. A Cheshire cheese weighing 1,235 pounds and using the milk of 900 cows was made in Cheshire, Massachusetts, sent by sleigh for 500 miles to Washington, D.C., and presented to President Thomas Jefferson on January 1, 1802. Inscribed with "Rebellion to tyrants is obedience to God," it was called the Mammoth Cheese due to the discovery of mammoth (now known to be mastodon) bones the year before. Cautious of accepting free gifts, Jefferson paid $200 for it and displayed it in the East Room for a year. Not to be outdone, supporters of President Andrew Jackson made a 1,400-pound cheese for him in 1837. But the largest cheese prior to the twentieth century was a 22,000-pound Cheddar made in Perth, Ontario, for the 1893 Columbian Exposition in Chicago. "The Canadian Mite" was made from 27,000 gallons of milk from 10,000 cows that were milked by more than 1,600 "maids." The curd came from 12 factories over a three-day period and was emptied into a hoop at the local railway station. The curd was pressed with 200 tons of pressure after each filling, forming a cheese 9 feet in diameter and 6 feet high.

A larger cheese was made for the Wisconsin Pavilion at the 1964–65 World's Fair in New York. The 34,591-pound Cheddar required 42,500 gallons of milk from 16,000 cows. Twenty men worked around the clock for 43 hours to make the cheese, which measured 14½ feet by 6½ feet by 5½ feet. The current record-holder is a 56,850-pound Cheddar made in Oregon by the Federation of American Cheese-makers in 1989.

TABLE 2.2

Comparison of classes of cheese

Class (key varieties)	Noted characteristic	Moisture (%)	Calcium (%)	pH
Fresh soft (Cottage)	Curd washed and "dressed"	80	0.08	5.0
Whey (Ricotta)	Coagulated whey protein	72	0.21	5.9
Pickled (Feta)	Stored in brine	55	0.50	4.2
Stretched curd (Mozzarella, Provolone)	Kneaded and stretched in hot water	44	0.74	5.3
Surface mold (Brie, Camembert)	*Penicillium camemberti* on surface	50	0.28	6.9
Smear-ripened (Limburger, Brick)	*Brevibacterium linens* brushed on surface	45	0.59	6.6
Interior mold (Blue cheese, Stilton)	Mold powder added to curd and by skewering	39	0.66	6.5
Cheddared (Cheddar, Cheshire)	Curds stacked and flipped to remove whey	37	0.68	5.4
Stirred curd (Gouda, Colby)	Whey rinsed out with water	40	0.71	5.8
With eyes (Emmental, Gruyère)	Propionibacteria added, CO_2 formed	35	0.90	5.6
Very hard (Parmesan, Romano)	Low moisture, often grated	30	1.12	5.4

Some Spanish varieties have unusual shapes. Valencia's Queso Servilleta ("towel cheese") is tied in a towel knotted at the top to drain the whey, resulting in a distinctive indentation. Tronchón is dome-shaped with a deep dimple on top. Galicia's Cebreiro is made with a hoop that does not encompass all of the curd, causing some to bulge outward from the top, imparting a mushroom shape to the cheese. Another Galician cheese, San Simón, resembles the business end of a torpedo. Several varieties such as Queso Panela are molded in a basket and take on that shape. The size of cheese also depends on tradition, as well as convenience and ease of shipping. Some huge cheeses have been made from time to time (See Box 2.12).

Cheese Characteristics

A list of classes of cheese along with common varieties and some compositional data is shown in Table 2.2:

Casein micelles absorb the lactic acid that is generated, driving out CCP, which dissolves in the whey. In general, cheeses that have less moisture have less lactose and less acid formation, and therefore higher calcium and pH levels. The pH levels of interior mold, surface mold, and smear-ripened cheeses go from 4.6 to near 7.0 because the yeasts and molds quickly metabolize the lactic acid and some of the protein, forming ammonia from the latter. The calcium in the micelles becomes insoluble and migrates to the surface in Brie and Camembert, softening the inside of the cheese.

The procedures for manufacturing each class of cheese are different. We will cover these in Chapters 4–14.

Age is something that doesn't matter, unless you are a cheese.

—Variously attributed to Mexican filmmaker Luis Buñuel (1900–1983) and American actresses Billie Burke (1884–1970) and Helen Hayes (1900–1993)

3

YOU'RE NOT GETTING OLDER, YOU'RE GETTING BETTER

Aging Cheese

AGE DOES INDEED matter with cheese. Enzymes continue to degrade the compounds in cheese during storage, generating a huge array of breakdown products that contribute to flavor and aroma. Moreover, the feed the animals ate, the climate they lived in, and their species all have a say in the cheese. Processing parameters are also important. Strictly speaking, *ripening* refers to the chemical and physical changes occurring in cheese during storage, and *curing* refers to humidity, temperature, and other factors, though the two terms are often used interchangeably. Aging cheese is usually held in a room with temperature and humidity control such as the one shown in Figure 3.1, though many artisanal cheeses are stored in a cave. Cheese caves may be naturally occurring, or they may be carved out of a hillside or dug out of the ground. Ventilation is a must for aging rooms and caves so that the gases given off during ripening are carried away. Vacuum-packaged cheeses are an exception since the product is not exposed to air.

Remember that the yield of cheese during manufacture is around 10 percent, with most of the water being lost in the process. Aside from water, cheese consists of protein, fat, carbohydrates, minerals, and flavor compounds in a form much more concentrated than in milk. A seemingly minor difference in milk can amount to a major one in cheese. In fact, many factors can cause a batch of cheese to turn out differently than expected (Box 3.1).

FIGURE 3.1 Aging room.
Source: Author's collection.

BOX 3.1
WHAT CAN GO WRONG?

"It's not really a cheese for eating—it's more for encasing in concrete and dumping in the ocean a long way from civilization."
—JASPER FFORDE, *Thursday Next: First Among Sequels*, 2007

Any number of factors can lead to a less-than-ideal cheese. When a cheesemaker sees that things are going wrong, some alterations in the procedure are made to increase the chances that the batch, or future ones, won't have to be discarded. The following table (B3.1) shows common problems for hard and medium-hard cheeses and the some of the procedures that may be used to fix them. The steps in italics are usually the most effective.

Other problems can also occur, such as equipment failures and power outages. So can natural disasters: on May 20, 2012, a strong earthquake and two aftershocks struck the region of Italy that produces Parmigiano Reggiano, the name-protected version of Parmesan. A total of 633,700 wheels were being aged at the time, and some 300,000 of them in twelve facilities were damaged when they tumbled onto the floor after the shelves fell over or collapsed. The wheels that were aged more than a year were chopped into pieces and sold. But damaged wheels less than a year old had to be sent away without being labeled as genuine Parmigiano Reggiano. They found use as ingredients in other products, but the financial loss from the inability to use the official designation amounted to millions of dollars.

BOX 3.1 *(continued)*

TABLE B3.1

Cheesemaking problems and solutions

Problem	Solution	Steps to take
Cheese is too soft	Increase expulsion of whey	Raise calcium content by adding calcium chloride to the milk Cut curd into smaller cubes *Raise cooking temperature* *Maintain temperature during final stirring* Raise pressing time or temperature
Cheese is too hard	Decrease expulsion of whey	Add water to milk or during cooking Cut curd into larger cubes *Lower cooking temperature* *Cool curd cubes for a few minutes before stirring ends* Add more NaCl to whey, causing curd cubes to swell Reduce pressing time or temperature
Cheese is too acidic	Decrease acid production	Reduce amount of starter culture *Add water to milk or add more water during cooking* Raise cooking temperature or start cooking early to slow lactic acid development
Cheese is not acidic enough	Increase acid production	Raise amount of starter culture Add less water *Reduce cooking temperature* Cool curd cubes for a few minutes before stirring ends
Insufficient microbial activity during ripening	Increase microbial growth	Increase amount of starter culture *Increase storage temperature* Increase storage humidity

Affinage

The aging process in cheese is sometimes called *affinage* (French for "refinement" and pronounced AH-fee-nahj), and a specialist in charge of aging cheeses to create the best possible product is an *affineur*. In some places, cheese manufacturers sell their unripened product to *affineurs* who have their own aging facilities. The *affineur* is the person who flips the cheeses to retain their shape, applies or wipes off mold, monitors rind development, and otherwise controls the development of flavor and texture. Maturing cheese is sniffed and visually inspected by inserting a *trier* (also called a *cheese iron*, shown in Figure 3.3) and extracting a long cylinder of a sample. The *affineur* breaks off a small piece to examine the texture, places the rest of the cylinder back in the cheese, and makes judgments about adjusting humidity and temperature and when the cheese is ready for sale. Not everyone agrees with the idea of paying an *affineur* to look after their cheese, however. Some cheese sellers feel that an *affineur* is an unnecessary expense that can be avoided by simply keeping an eye on the product. A *New York Times* article presented both sides of the issue, followed by the results of a blind taste test of three cheeses. The samples from the one store that did not have an *affineur* scored much worse than the cheeses from two stores that did employ one.

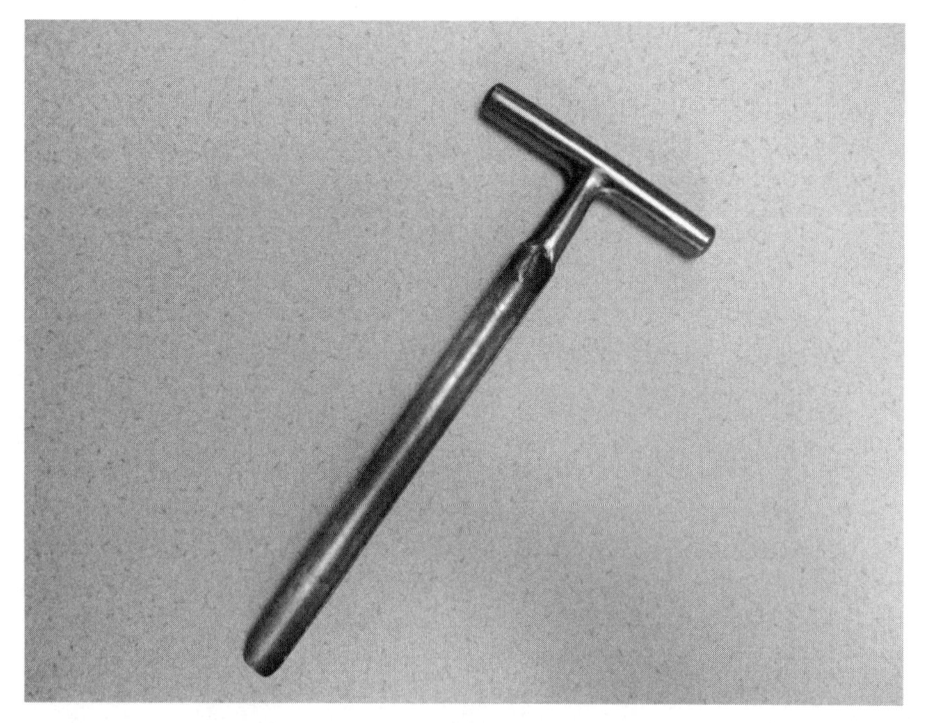

FIGURE 3.2 A trier is pushed or twisted into a block of cheese to extract a sample.
Source: Author's collection.

Length of Time

The number of weeks or months a cheese is aged depends on the variety and what the cheesemaker, *affineur*, wholesaler, or retailer wishes to accomplish. The longer a cheese is allowed to sit before it is offered for sale, the more flavor compounds are formed, and the more the texture changes. If a milder cheese is desired, the aging process will not be nearly so long as that for a sharper, more flavorful cheese. The cheese has to be sold before undesirable flavors crop up or the body softens too much. Some of the cheeses that are lower in moisture can be aged for years. Some Cheddars that were unknowingly aged in wooden boxes for 28, 34, 39, and 40 years were discovered in 2012 in the back of a cooler and were found to be very sharp but creamy and not dry. The oldest cheese was sold for $10 an ounce.

Feed and Species

Milk from animals grazing on pasture is not the same as milk obtained from animals eating conventional feed, and the differences become noticeable as a cheese ages. American cows are commonly given *fodder*, which includes hay, silage, and total mixed ration (more on this in Chapter 11). Pasture-fed cows, on the other hand, consume *forage*—they graze and eat what they want. Some of the compounds from flowers, grass, and weeds either pass directly into milk or are converted in the animal into other compounds. Many of these compounds impart desirable flowery flavors to cheese.

As we saw in Chapter 1, the composition of milk varies among species. The proteins and fats in the milk are also different. Box 2.7 listed the four types of casein in milk: α_{s1}-, α_{s2}-, β-, and κ-casein. Goat's milk contains little or no α_{s1}-casein, which is the primary structural protein in cheese. Soft goat's milk cheese is quite flimsy; if you start to unwrap a goat cheese that was shaped into a log, one end will fall off under its own weight. Goat's and sheep's milks contain fatty acids in different proportions than cow's milk, resulting in more piquant aged cheeses. The influence of feed and species on cheese flavor is discussed in Chapter 5.

Terroir

Terroir (from the French for "land" and approximately pronounced terr-WAHR) is the word that encompasses altitude, soil, and weather on a farm and their effects on the food it produces; especially wine, but including cheese. Consumers' attitudes toward a sense of place carry over to the food coming from there, especially if a small-scale operation is involved. *Terroir* implies authenticity, culture, history, know-how, and reputation, making it a desirable characteristic for artisanal cheeses. The effects of *terroir* on cheese as it ages are described in Chapter 13.

Processing

Pasteurization inactivates several milk enzymes and kills nearly all of the indigenous microorganisms as well as bacteria that enter the milk from the environment. The resulting cheese therefore contains fewer flavor compounds than raw-milk cheese because of decreased protein and fat breakdown. For example, a study of Cheddar aged for eight to 24 weeks showed that Cheddar made from pasteurized milk contains 25–30 percent fewer fatty acids than raw milk cheese. Although protein breakdown is unaffected, pasteurized-milk Cheddar has less intense flavor since fewer fatty acids are available to degrade into flavor compounds. Another study indicated that French cheeses made from raw milk have more aroma and somewhat different flavors than those made with pasteurized milk. Some enzymes such as plasmin do survive pasteurization, however, and are present in the milk during cheesemaking.

The temperature of cooking, the size of the curd pieces, the amount of pressure used, the storage conditions, and a host of other processing factors all contribute to the quality of the final product—and the product isn't really final, since biochemical processes are taking place until the last piece is eaten.

Microbes

The lactic acid bacteria in the starter culture mostly die off while cheese ages, and often their cells open up and spill out their contents, including enzymes responsible for the formation of flavor and texture. If these microbes stay alive, the cheese won't ripen as quickly. In many varieties, Lactobacilli that are added or come in from the environment begin to dominate as Lactococci disappear. Most of the remaining lactose is metabolized by surviving microbes.

Sometimes mold will grow on the outside of a piece of cheese after you have brought it home. This mold is not part of the aging process—it comes from spores in your refrigerator and was not in the cheese when you bought it. If it is a hard cheese, you can cut off the portion containing the foreign mold and eat the rest. Mold can penetrate softer cheeses and send veins into the interior, so these varieties must be discarded if you notice contamination.

Flavor Compounds

How many flavor compounds are in cheese? One study in 2004 looked at three Cheddars, one medium-flavored blue cheese, two strong blue cheeses, and three very hard cheeses. The researchers found 73 odor-active compounds, of which 61 could be identified. Only four compounds were located in all nine of the samples they tested, which highlights the diversity of flavors in the world's cheeses. According to Professor Robert McGorrin of Oregon State University (and formerly of Kraft Foods), Cheddar cheese has from 750 to 1,000 volatile compounds, with about 250 of them being significant contributors to flavor.

The major flavor compounds in cheese may be grouped into eleven categories:

- Acids
- Amino acids
- Fatty acids
- Alcohols
- Sulfur-containing compounds
- Aldehydes
- Ketones
- Esters
- Lactones
- Furans
- Terpenes

The basic structures of most of these are compared in Box 3.2.

Acids are substances that, when dissolved in water, lower the pH. These include mineral acids (such as sulfuric acid), nucleic acids (DNA), and carboxylic acids (which have the COOH group we saw in Box 1.1). Carboxylic acids are the only ones found in cheese at appreciable concentrations, and for our purposes we will classify them into acids, fatty acids, and amino acids.

Amino acids are the links that, when assembled properly, make up a protein. They each contain a nitrogen atom, and we will discuss them along with other nitrogen-containing compounds.

Fatty acids are breakdown products of triglycerides and are precursors to many flavor compounds.

Alcohols are familiar molecules. The smallest is methanol (wood alcohol), with one carbon atom, and the next is ethanol (grain alcohol) with two carbons. The alcohols responsible for flavor and aroma in cheese have longer chains.

Sulfur-containing compounds include methional, methanethiol, and various sulfides.

BOX 3.2
STRUCTURES

We will use a chain of eight carbon atoms to illustrate the differences between compounds. Remember that each end and bend represents a carbon atom unless marked otherwise.

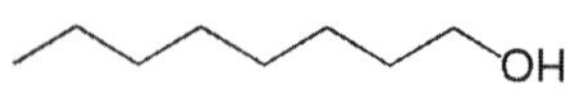

Alcohols have names ending in *ol*. The molecule above is called octanol since it has eight carbon atoms and *oct* is the prefix for eight. Alcohols are characterized by the OH attached to a carbon atom, representing an oxygen atom bonded to a hydrogen atom.

Aldehyde names end in *al*. This example is octanal, which has an oxygen atom double-bonded to the carbon atom at the end.

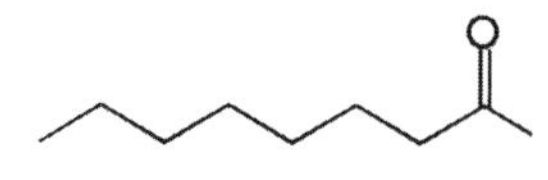

Ketones look like aldehydes, but have the double-bonded O attached to two carbons somewhere in the middle. This molecule is 2-octanone, where *2-* designates which carbon has the oxygen, and *-one* denotes a ketone.

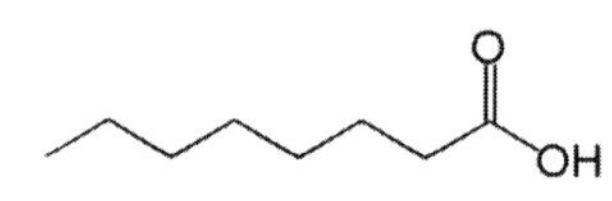

The above is caprylic acid, also known as octanoic acid. It is classified as a fatty acid since it is derived from fat. Note the COOH group at the end.

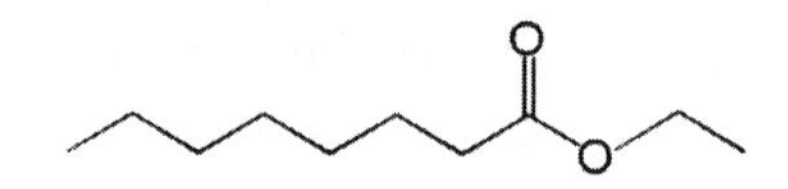

Esters are combinations of acids and alcohols. This example (above) is ethyl octanoate, where *ethyl* denotes the two carbons on one side of the oxygen, *octan* is for the eight carbons on the other side, and *-oate* means an ester. Had there been one carbon on the end instead of two, the compound would have been methyl octanoate. *Methyl* denotes a single carbon atom attached somewhere on a chain.

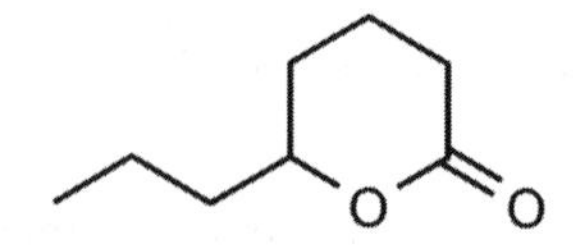

Lactones are a category of esters known as cyclic esters, with a multi-sided ring of atoms. The molecule above has eight carbon atoms—five in the ring and three on a "tail." This atom is δ-octalactone, where δ (delta, the fourth letter of the Greek

BOX 3.2 (*continued*)

alphabet) means that five carbon atoms are in the ring. Lactone comes from the Latin *lac*, "milk," also the source of the words lactose and lactic. An important lactone in cheesemaking is glucono-delta-lactone, a common food additive which the body metabolizes to glucose. It has this structure:

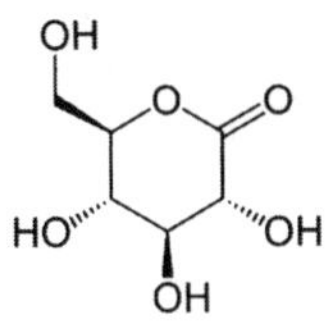

The thick and dashed lines are for visualizing the molecule in three dimensions. The parts of the molecule with the thick lines can be imagined as coming up from the paper and pointing toward you. The parts with the dashed lines go through the paper and are pointed away from you.

The structures of amino acids, sulfur-containing compounds, furans, and terpenes will be covered in the upcoming chapters.

Aldehydes, like alcohols, are relatively simple. The smallest is formaldehyde, solutions of which are used as embalming fluid and fortunately not found in cheese. Aldehydes are rapidly converted into alcohols or acids.

Ketones are structurally similar to aldehydes but are more plentiful in cheese. The simplest ketone is acetone, used as nail polish remover, and a tiny amount of it is found in cheese.

Esters are important flavor compounds in many fruits and fruit products. They have low perception thresholds, which mean that a little goes a long way. The aroma of isoamyl acetate can be detected when it is dissolved in water at a concentration of 2 parts per billion.

Lactones are also detectable at the parts-per-billion level. They have properties like esters since they are cyclic versions of them, meaning that some of the carbon atoms are arranged in a closed ring instead of an open chain.

Furans are derived from breakdown of carbohydrates. The furans in cheese primarily consist of furaneol, homofuraneol, and sotolon.

Terpenes come from pasture plants. Terpenes are not found in milk or cheese from cows on conventional feed.

In the following chapters, we will explore the characteristics of the flavor categories and pair them off with the classes of cheese. Cheese texture also develops during aging, and that subject will be covered in Chapter 10.

The cottage cheese has doubtless softened his brain.

—MIGUEL CERVANTES, *Don Quixote*, 1605

4

FRESH SOFT CHEESES, ACIDS, AND SAFETY

THE TWO MOST eaten fresh cheese types in the United States are cottage cheese and cream cheese, and the most popular fresh varieties in Latin America are Queso Blanco and Queso Fresco. These cheeses have relatively little flavor since they are not aged and there is not much time for flavor compounds to form. Nevertheless, they do have some flavor, and we will explain how it arises.

In this and the following ten chapters, the procedures for making popular varieties will be described, along with the formation of their characteristic flavors and related topics. Note that scientists have not found all of the pathways that create flavor compounds, nor do they agree on what some of them actually taste like in cheese. As we will see, the concentration of a substance and the other molecules present can mean a great deal.

Cottage Cheese

Cottage cheese, cream cheese, and Queso Blanco are *acid-coagulated varieties*, meaning that little or no rennet is added. In this aspect, they are similar to yogurt (Box 4.1). The manufacture of cottage cheese, which got its name because it was made in small country homes from leftover milk, begins when starter and possibly some rennet are added to milk at 70°. For "short set" cottage cheese, the curd is held at 88° for 5 hours; for "medium set," 83° for 8–9 hours; for "long set," 72° for 12–16 hours. Cheese made without rennet is cut at pH 4.8 with quarter-inch knives, resulting in small-curd cheese. When made with rennet, the curd is cut at pH 4.8 with 3/8-inch

BOX 4.1
THE DIFFERENCE BETWEEN CHEESE AND YOGURT

Like cheese, yogurt ("to thicken" in Turkish) is produced using starter cultures, usually *Lactobacillus bulgaricus* and *Streptococcus thermophilus* and sometimes including bifidobacteria and *Lactobacillus acidophilus*. Unlike cheese, the milk is heated to 185° for 30 minutes or 200° for 10 minutes, which kills off any pathogens and causes the milk protein to denature, or fall apart. By doing this, the milk thickens instead of forming curds. The starter is added after the milk is homogenized and cooled to 110°, and the temperature is held there for 2–3 hours. The result is a gel with a pH of 4.5. No effort is made to remove whey, so fruit and flavor are often added to neutralize some of the tartness.

Yogurt production was exclusively artisanal until 1919, when Isaac Carasso, whose oldest son was named Daniel, began industrial manufacture of yogurt in Barcelona, Spain. He named the company *Danone*, Catalan for "little Daniel." The son took over the company in the late 1930s, moved it to New York in 1942, and changed the American version of its name to Dannon.

Strained (or Greek) yogurt is strained to remove whey, producing a thicker product. A pound of milk is needed to make a pound of regular yogurt, but four pounds are required to obtain a pound of strained yogurt. *Labneh* (from a Semitic word for "white") is a similar product popular in the eastern Mediterranean.

knives to create large-curd cheese. Cutting must be gentle so that the fragile curd doesn't shatter into tiny pieces, known as *fines* (which would be lost with the whey, decreasing the yield). The curd is allowed to heal undisturbed for 15–30 minutes, which allows the curd pieces to shrink and expel some whey. The curd is cooked to 117°–133° over 1–3 hours, which kills undesirable bacteria and causes the curd to become firmer as more whey is driven off. Then the curd is washed three times with water at 72°–90°, 50°–61°, and 36°–41°, removing lactic acid and lactose and preventing additional acid production. The washing step also cools the curd pieces and keeps them from clumping together. After the water is drained, the curd is dressed by adding cream containing small amounts of salt and stabilizers such as guar gum or xanthan gum (see Box 4.2). Dry curd cottage cheese, also called *pot cheese* in England, is made the same way but is not dressed. Cottage cheese is not pressed; *farmer's cheese* is the pressed version, and is found in sliceable loaves or logs.

An alternate method of making cottage cheese is by adding acid to cold milk with no starter or rennet. The milk is slowly warmed to 90° and glucono-delta-lactone is added. This lactone converts to gluconic acid in water, which lowers the pH and causes the CCP to dissolve and the casein micelles to aggregate into a network. The pH reaches 4.5–4.7 within an hour, and the product, labeled "direct set," is packaged.

BOX 4.2
STABILIZERS

Xanthan gum, guar gum, and locust bean gum (also called carob bean gum) are polysaccharides, a carbohydrate in which a section of the molecule repeats itself many times to form a long chain. Xanthan gum is isolated from the coat of the bacterium *Xanthomonas campestris*, which may be grown in whey, where it metabolizes lactose. Guar gum is extracted from guar beans, a legume grown mostly in India. Locust bean gum comes from seeds of the carob tree, found in the Mediterranean region. These gums are effective thickeners—a small amount greatly increases the viscosity of a liquid (water has low viscosity and ketchup has high viscosity). They also exhibit shear thinning, which means that chewing or mixing it lowers its viscosity, and when the chewing or mixing stops, the viscosity recovers. They are used in salad dressings for this reason since it causes the liquid to flow easily when the shaken and poured, and then thickens again and sticks to the salad. Xanthan and guar gums control the viscosity of cottage cheese dressing. All three gums stabilize cottage and cream cheese by coating the fat droplets and keeping them from separating out.

Quark

When milk is soured by allowing the starter bacteria to grow for 14–18 hours, heated to 200° so the casein falls apart, and then strained, the result is a fresh cheese known in Germany as *quark* or *quarg* (from *twarc*, "curd"), in Russia as *tvorog* (same derivation), in Austria as *topfen* ("pot cheese"), and in France and other places as the local translation of "white cheese." Like cottage cheese, it is often made without rennet. Cream may be removed or added, resulting in a cheese with anywhere from zero to 40 percent fat. Quark normally has a creamy texture, but adding a small amount of rennet leads to a firmer cheese.

Cream Cheese

Cream cheese is another fresh, soft, acid-coagulated cheese. Homogenized milk is used to reduce fat loss (the fat globules become coated with casein) and to produce a soft, smooth curd and a white appearance. Using only a starter, the curd is held at 72° until the pH reaches 4.5. Agitators break the curd, which is then heated to 175°–185° to help syneresis and inactivate the starter bacteria. The curd is not cut, and about 0.75 percent salt is added, along with 0.25 percent guar gum or locust bean gum to stabilize the cheese. After homogenization to mix the ingredients, the cheese is packaged without cooling ("hot-packed") or after chilling to room temperature ("cold-packed"). The soft American cheese labeled as Neufchâtel is made the same way but contains 20–33 percent fat, whereas cream cheese is required to contain more. Cream and Neufchâtel cheeses have smooth textures with no curd pieces visible. Note that American Neufchâtel is not related to the French cheese of the same name.

European Varieties

Brin d'Amour (French for "bit of love") is a fresh sheep's milk cheese made in Corsica and also known as Brindamour and Fleur du Maquis ("flower of the macchia," Mediterranean shrubland). The 5-inch squares are covered with blue-gray mold and rosemary and other local spices. It is soft after a week, almost runny after a month because of proteolysis, and hard after two months as it dries out. The flavor intensifies with time as compounds are absorbed from the exterior.

Crowdie, from Scotland, is also similar to cottage cheese. An artisanal British variety, Caboc, has a crunchy exterior and an interior with a texture similar to that of cream cheese. According to legend, this variety dates back to the fifteenth century when a noblewoman fled from the Isle of Skye to Ireland to avoid abduction and marriage; she brought the recipe for Caboc with her when she returned. The variety was revived in 1962 by a descendant of the creator. Starter cultures are added to cream-enriched milk, which is allowed to stand in milk cans for three months. The soft curd is drained in bags, shaped into logs, and rolled with toasted oatmeal, which does not contribute to flavor.

The Italian soft cheese Bel Paese (bell pah-AY-zeh, "beautiful country") was invented in 1906 and is uncooked, pressed, and coated with wax. Also made in Wisconsin, it is ripened for a month or two. Mascarpone (ma-skar-POH-nay) is not considered by many to be a cheese since starter and rennet are not used to produce it. Warmed cream is coagulated by citric acid, tartaric acid, or vinegar, and is drained through cheesecloth or, in large-scale operations, separated by centrifugation. The product resembles dense cream with a fat content over 70 percent. A common use is in the dessert tiramisu (Italian dialect for "pick-me-up"). Mascarpone dates back to the sixteenth century, and the origin of the name is debated. It could have come from a visiting Spanish official who called it *mas que bueno* ("better than good"), or it was derived from a Ricotta named Mascherpin, or it comes from *mascarpa* ("whey remaining from cheese manufacture"), or it was derived from *mascarpia* (dialect for Ricotta), or it was named after a dairy farm that made it, or it was derived from *mascherare*, the Italian verb for "dress up," referring to the treatment of the cream.

Queso Blanco and Queso Fresco

Queso Blanco and Queso Fresco are bright white unaged cheeses that originated in Spain and spread to Central and South America. Queso Blanco is Spanish for "white cheese" and is made by heating milk to 185° and adding acid such as vinegar. The casein and whey proteins protein coagulate, the whey is drained, and the curd is dry-salted, packed in hoops, and pressed. Paneer is a similar cheese from India, but little salt is added. Chhena, from eastern India and Bangladesh, is kneaded instead of pressed. Quark is a German and Eastern European variety in which acid-coagulated milk is stirred to produce a thick and creamy texture, and then strained.

Spanish for "fresh cheese," Queso Fresco differs from Queso Blanco because it is coagulated with rennet, at 86°–90°. The curd is cut after 45 minutes, cooked at 115°, and stirred. The whey is drained after an hour, and the curd is milled into fine pieces with a grinder. The curd is then dry-salted and packed into forms, shown in Figure 4.1. Queso Fresco is a non-melting cheese used for frying. It doesn't melt, because its pH is about the same as milk's (6.5–6.7), which allows the CCP to remain in the protein matrix and hold it together. Melting is also impeded because the milk is homogenized, which reduces the size of the fat globules and allows them to be held tight in the protein matrix. The milling step also makes it a crumbly cheese.

FIGURE 4.1 Queso Fresco. The crumbly nature of this cheese is evident from the surfaces of these cut pieces.

Source: Author's collection.

Unfortunately, Queso Blanco and Queso Fresco have been implicated in outbreaks of food-borne illness. Many health professionals recommend that pregnant women avoid fresh Hispanic cheeses for this reason. In truth, any cheese ought to be safe as long as it is made under sanitary conditions.

Safety

The safety of a food is of paramount importance. Its enjoyment is of no value if the consumer becomes sick from eating it. Cheese that is made without sanitation in mind is at risk for food-borne illness since it provides such a good environment for microbial growth. Microbial contamination of commercial cheese is rare, however, because manufacturers implement food safety plans. A prominent system used by the food industry is Hazard Analysis and Critical Control Points (HACCP), which was originally designed in the 1960s for providing contaminant-free food for astronauts. HACCP has seven parts:

- Conduct a hazard analysis, identifying all safety hazards and measures that should be taken to control them.
- Locate critical control points where a preventive measure can be applied.
- Set a limit that the critical control point cannot exceed.
- Set the monitoring requirements for each critical control point.
- Establish corrective actions to be taken if a limit is exceeded.
- Set a procedure to insure that the HACCP system is working properly.
- Keep records.

There have been no outbreaks of illness resulting from cheese made in plants using HACCP or any effective food safety system. Unfortunately, some people are involved in making illicit "bathtub cheese," which is homemade cheese made under less than sanitary conditions and then sold door-to-door, from the back of a vehicle, etc. Queso Blanco and Queso Fresco appear to be the two most common cheeses made in this fashion. It is important for consumers to purchase labeled cheese from an established store.

Inspections

One way to insure that a cheesemaking operation is following good hygienic practices is by having an official inspection. The Dairy Grading Branch of USDA's Agricultural Marketing Service inspects about 150 cheese plants at least twice a year. Inspections are voluntary, unannounced, and cover more than 100 items, including milk supply, plant facilities, condition of equipment, sanitary practices, and processing procedures. The instruction book for inspectors is over 400 pages long and includes butter, cheese, dried products, frozen desserts, ice cream, milk, whey, and yogurt. When inspection is completed, the inspector meets with the plant manager to review any steps that must be taken before approval can be granted. After corrections are made, the inspector comes back, checks the place again, and, if all goes well, approves it. If a facility is approved, its products become eligible for grading, which is an impartial opinion of the quality of the product. Grading is discussed in Chapter 15.

Pathogens

The most dangerous pathogen in the cheese industry is *Listeria monocytogenes*, the source of listeriosis, which is often fatal. This bacterium lives in soil, decaying vegetation, water, and feces, but cannot survive pasteurization. It has been traced to floor drains and other places in food plants with standing water, giving it the potential to contaminate cheese after pasteurization. *Listeria* is a hardy bacterium since it grows even under refrigeration and persists for months. Queso Blanco, Queso Fresco, and other soft cheeses have high pH and moisture levels, making them particularly susceptible to contamination by *Listeria*, which may thrive in cheese with a pH over 6 and moisture content over 50 percent. The largest listeriosis outbreak in U.S. history occurred in mid-1985, when 48 died and nearly 100 others became ill from consuming soft Hispanic-style cheeses made from unpasteurized milk. A soft smear-ripened cheese, Vacherin Mont d'Or, caused over 120 cases of listeriosis in Switzerland between 1983 and 1987, resulting in 34 deaths. That cheese was made from pasteurized milk, and the *Listeria* was traced to contaminated brushes and wooden shelves. Recalls of soft cheeses such as Brie and smear-ripened cheeses such as Leiderkrantz and Limburger have taken place in the United States and Europe because of *Listeria* contamination. *Escherichia coli* and species of *Campylobacter* and *Salmonella* have been implicated in many outbreaks of food-borne illness, including a few in cheese. These have been traced to improperly pasteurized milk or post-pasteurization contamination. Like *Listeria*, these organisms do not survive pasteurization. Unlike with *Listeria*, people rarely die from being infected with these bacteria. Figure 4.2 shows what *Listeria* and *Salmonella* look like in cheese under an electron microscope (see Chapter 15 for how an electron microscope works). Pasteurizing milk and conducting a proper sanitation program are the best ways to prevent problems from dangerous bacteria in cheese. Cheese made from raw milk is also safe as long as it is free of pathogens. The Food and Drug Administration requires aging for at least 60

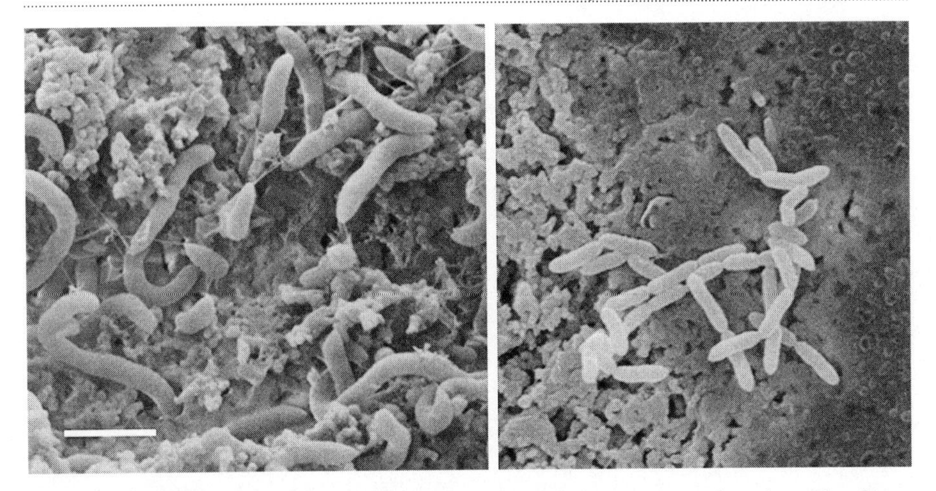

FIGURE 4.2 *Listeria* (left) and *Salmonella* (right) in Queso Fresco samples made and inoculated in our laboratory. The experiments were performed to develop new ways of killing these bacteria. The white line at lower left corresponds to 2 micrometers.
Source: Author's collection.

days at 35° or higher for any raw-milk cheese, with the assumption that any pathogens present should die out because of the high NaCl and low pH and water content. The FDA is taking another look at the regulations since research has shown that some pathogens are able to survive the 60-day period, while others will not survive in cheeses stored between 40 and 86° and containing traditional pH and salt levels and less than 50 percent moisture. A recent study of 41 raw milk cheeses from across the United States indicated that the FDA rules are adequate: no pathogens were seen. The shelf life of fresh cheese is less than 60 days, meaning that raw milk cannot be used for those varieties.

None of the above implies that cheese is dangerous to eat. If you stick to commercial cheese sold at your local store, the chances of getting sick from eating it are essentially nil.

Acids

The cheeses described in this chapter are not aged but do have flavor. The sources of flavor compounds in cheese are lactic acid, citric acid, proteins, and lipids, all of which are degraded into simpler molecules. Salt also provides flavor, but it remains intact.

Lactic and citric acids are present in milk, and formic and acetic acids are among their breakdown products. Formic acid, from the Latin *formica* (ant, since it is found in red ant venom) has one carbon atom, and acetic acid, from the Latin *acetum* (vinegar,

BOX 4.3

LACTIC AND CITRIC ACID METABOLISM

We saw the structures of lactic acid and citric acid in Box 2.6 and the structure of glucose in Box 1.7. All are metabolized by bacteria in a series of reactions to form compounds such as pyruvic acid:

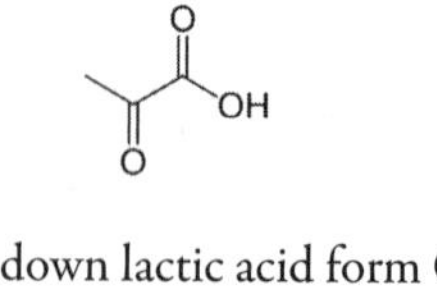

Enzymes in bacteria that break down lactic acid form CO_2 as well as—

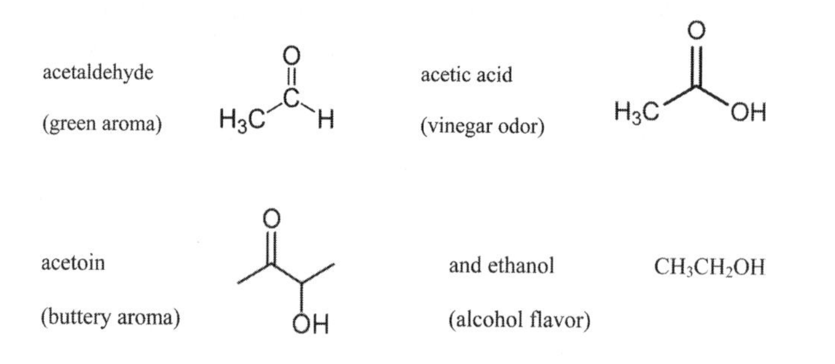

The "green aroma" of acetaldehyde refers to the scent of cut grass. The bacteria that break down citric acid form CO_2, acetaldehyde, acetic acid, acetoin, and probably the most flavor important compound in this group:

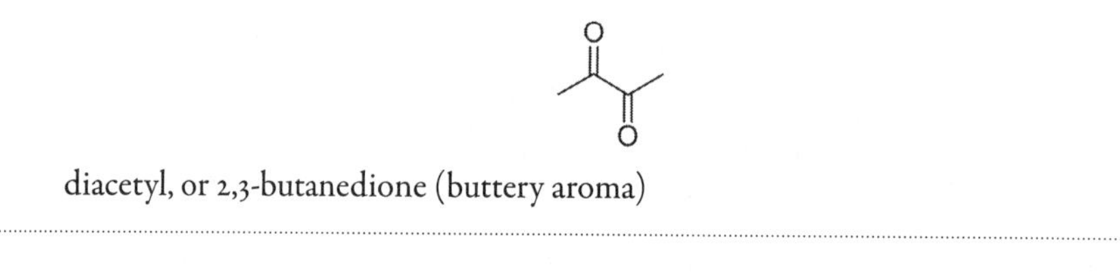

diacetyl, or 2,3-butanedione (buttery aroma)

which is diluted acetic acid), has two carbon atoms. Propionic acid has three carbons and a pungent, vinegar flavor similar to acetic acid's. The name comes from the Greek *pro* + *pion* (forward + fat, because it is at the top of the list of compounds that are similar to fatty acids). "Propane" is derived from this word.

These compounds impart a pungent acid flavor. The lactic acid that is not metabolized by bacteria gives cottage cheese its acidic taste, and is present at up to 450 parts per million. Lactic acid is not volatile, but formic acid and acetic acid, which are both present at up to 300 ppm, are volatile and are partly responsible for the aroma.

Cottage cheese contains over 2.5 percent lactose, and cream cheese has over 3 percent, whereas an aged cheese has almost none. The bacteria that convert lactose into lactic acid continue to work after the cheese is made, breaking down the lactic acid into pyruvic acid, which is the precursor to several flavor compounds (Box 4.3). Citric acid is also present in cheese; Cheddar contains about 0.2–0.5 percent. Mesophilic bacteria (but not thermophilic bacteria) metabolize citric acid to diacetyl, the primary flavor compound in cottage and cream cheese. Metabolism of lactic and citric acids is also responsible for CO_2 formation, which creates the eyes in Swiss and Dutch-type cheeses.

Cheese contains two other important classes of acid. The breakdown products of proteins include amino acids, and lipids are degraded into fatty acids. These are covered in the next chapter.

If you think cheese is just food, you're an eejit,
because it is so much more than that. It's poetry.
It's passion. It's pathos. It's no coincidence that
milk and human blood are almost the same
temperature. Had you thought about that, now?
—SARAH-KATE LYNCH, *Blessed Are the Cheesemakers*, 2002

5

WHEY AND PICKLED CHEESES, AMINO AND FATTY ACIDS, AND SALT

NOW WE GET into more flavorful cheeses and the factors that help make cheese more than "just food." Amino acids and fatty acids are the classes of molecules that are cloven into many of the compounds mentioned in Chapter 3. Salt also enhances flavor and helps control microbial growth.

Ricotta

Ricotta (Italian for "recooked") is the only common American cheese made from whey. Starter culture or acid is added to whey generated from production of another cheese variety, and once the pH reaches 5.6–6.0 the whey is heated to 185°. The whey proteins (the ones mentioned in Box 2.9) will collect on the surface over a 30-minute period. The curd is removed with a perforated scoop and filtered to concentrate the protein. Ricotta is not pressed. Two varieties similar to Ricotta are Requesón (made in Spain and Mexico) and Requeijão (Portugal and Brazil). A small Requesón operation is shown in Figure 5.1.

Gjetost or Geitost (Norwegian for "goat cheese" and pronounced YAY-toast) is cooked goats' milk, cream, and whey. The mixture is boiled for several hours, driving off most of the moisture and caramelizing the lactose. The cheese, which is not aged, is brown with a smooth and firm texture, and has a sweet caramel flavor. An example is shown in Figure 5.2. Mysost and Brunost are cows' milk versions.

FIGURE 5.1 Artisanal Requesón set-up in Mexico, with a propane tank for providing heat.
Source: Author's collection.

FIGURE 5.2 Gjetost.
Source: Author's collection.

Feta

Feta (from the Italian for "slice," since it does not fall apart when being sliced) is traditionally made in Greece from sheep's milk, with up to 30 percent goat's milk allowed. Over half of the cheese consumed in Greece is Feta. This variety is ancient, and Homer apparently wrote about it in *The Odyssey* in the eighth century BC (Polyphemus, the Cyclops who devoured six of Odysseus' men before they blinded him and escaped, kept sheep and goats and "dressed the milk up for cheese and pressed it into wicker cheese baskets"). Feta and other cheeses ripened in brine such as Domiati (which originated in Egypt), Halloumi (Cyprus), and Telemes (Romania) have high NaCl concentrations to allow for preservation without refrigeration (more on salt later). Domiati, in fact, is made with milk to which NaCl has been added, at concentrations up to 15 percent.

Milk is cooled to 93°–96°, Lactococcus and Lactobacillus starter cultures and lipase (an enzyme that breaks down fat) are added, and calf or lamb rennet is added after 15–30 minutes. An hour later, the curd is cut into three-quarter-inch cubes without cooking, scooped into perforated forms, and held for several hours at 60°–64° to drain the whey. Dry salt is placed between slabs of curd in plastic containers. After a day or two, the cheese is washed and repackaged into barrels or cans filled with brine containing 6–8 percent NaCl, and stored at 60°–64° for up to two weeks. The pH is around 4.5, making Feta one of the most acidic cheese varieties. Feta is frequently sold in crumbled form, as consumers often use it in that form in salads and other dishes.

Feta contains up to 100 ppm fatty acids, and nearly that quantity of acetic acid. Acetic acid is a major contributor to the flavor of Feta, which also contains alcohols, aldehydes, and ketones.

Bryndza

Bryndza (from the Romanian for *cheese*, though it is also slang for "poverty") was developed in eastern and central Europe and first mentioned in the late fifteenth century. Shepherds would add rennet to sheep's milk, drain the curd in cloth bags, and bring the curd to market each week. The curd was salted, pressed, and brined or packed into barrels with salt. Nowadays, the cheese is placed on shelves at 65°–68° and flipped regularly for two or three days, then transferred to another room at 59° for 3–6 days. The dried surface is scraped away, and the cheese is pressed for two days. Finally, the cheese is broken into pieces, possibly mixed with cow's milk cheese (it must contain at least 51 percent sheep's milk), salted to a concentration up to 2.5 percent, and pressed into a paste between rollers. Bryndza is similar to Feta, but softer and less salty.

Amino Acids

Protein consists of long chains of amino acids: β-casein has 209 amino acids, α_{s1}-casein has 199, and κ-casein has 169. There are 20 amino acids required for life, all are found in cheese, and some of them are shown in Box 5.1. Amino acids are not responsible for saltiness but do account for the other four basic tastes, as we will see in Chapter 7.

BOX 5.1
AMINO ACIDS

The building blocks of protein, amino acids also help build the flavor profile of cheese. Almost all amino acids have the same basic structure shown here, where R is the "side chain" that distinguishes them, and NH_2 is the *amino group.*

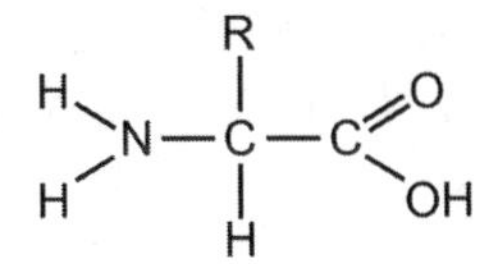

Microbes have several basic ways of breaking down amino acids. One is by using lyase enzymes to split off the side chain (lactic acid bacteria don't do this, but yeasts and molds do). Another involves aminotransferase enzymes, which transfer the amino group to another molecule. Aminotransferases are found in most cheese microbes. The ultimate breakdown product of the amino group is ammonia, NH_3, which has a familiar and unpleasant odor.

Following are some of the amino acids whose breakdown products lead to significant flavors.

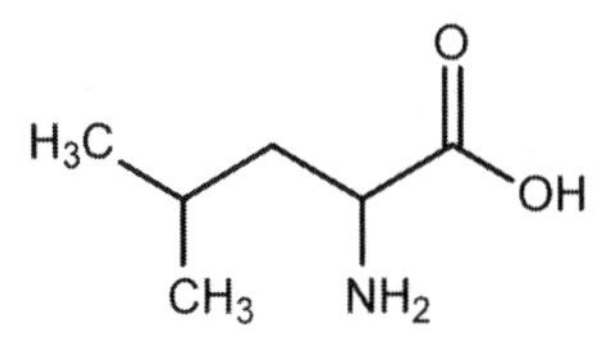

Leucine: Bacteria degrade branched-chain amino acids such as leucine into 3-methylbutanal, 3-methylbutanol, and 3-methylbutanoic acid. Isoleucine and valine, which are also branched, break down into similar compounds.

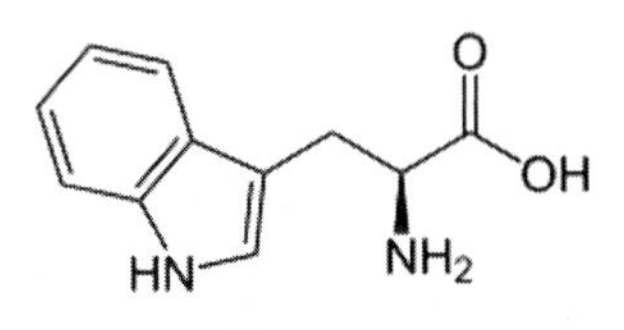

(*continued*)

BOX 5.1 *(continued)*

Tryptophan: Bacteria, yeast, and *Brevibacterium linens* are capable of splitting this molecule, producing indole and skatole (Box 5.2).

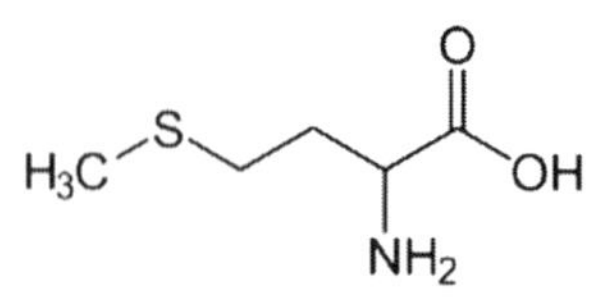

Methionine: The sulfur compounds in cheese mostly originate from this amino acid. Lactobacillus and Lactococcus bacteria, *Brevibacterium linens, Geotrichum candidum,* and *Penicillium camemberti* can all convert methionine to methanethiol, dimethyl disulfide, dimethyl trisulfide, and other compounds, which we will see in Chapter 7. Another sulfur-containing amino acid, cysteine, is found at low levels in casein, though it may also contribute to formation of these compounds.

Phenylalanine and tyrosine: Tyrosine is pictured below; phenylalanine is the same, but without the OH group attached to the ring. These amino acids lead to the production of aromatic compounds, including benzaldehyde, phenylacetaldehyde, 2-phenylethanol, and phenylacetic acid.

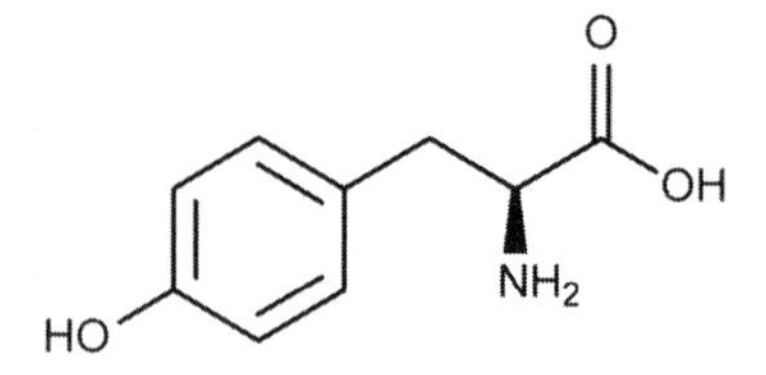

Two amino acids in cheese do not contribute to flavor but can be converted to compounds that cause discomfort. One product of tyrosine is tyramine, which causes headaches in some people. This "cheese effect" is observed in aged cheeses such as Cheddar or Stilton.

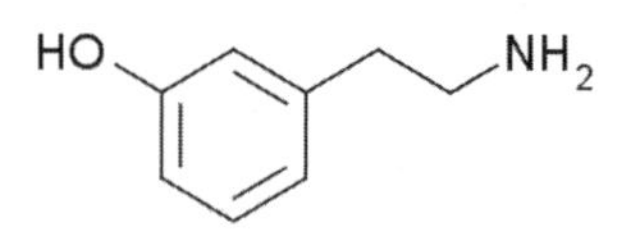

Another amino acid than can lead to trouble in susceptible people is histidine. Enzymes chop off the COOH group, converting the molecule into histamine, which produces the inflammatory response. People with allergies take antihistamines to inhibit this response.

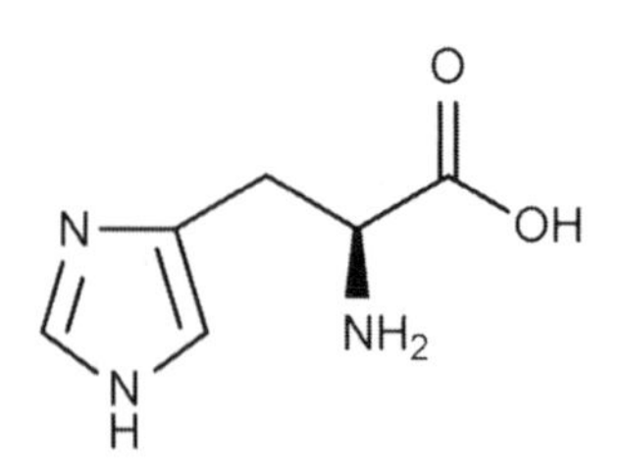

Molecules are three-dimensional, and their atoms can be oriented in different directions. In fact, amino acids and sugars (such as lactose) come in left-handed and right-handed forms. Your left hand is a mirror image of your right hand, but you can't get your thumbs and other fingers to coincide—one hand will always be aligned incorrectly in relation to the other. This "handedness" was discovered in molecules by Louis Pasteur in the mid-1800s. Virtually all amino acids are left-handed and virtually all sugars are right-handed, which enables them to react with each other, but scientists don't know why the handedness isn't the other way around. Biology as we know it would not be possible if amino acids were a mixture of left and right, so something drove them toward one orientation. Many believe that left-handedness of amino acids and right-handedness of sugars developed at random, but another answer may have come from outer space. A meteorite weighing over 220 pounds fell on Murchison, Victoria, Australia, in 1969, and has been found to contain dozens of amino acids. Many of them, such as isovaline, are rare or not found naturally on Earth. Only half of the 20 amino acids needed for life were found, which eliminates the possibility of contamination from terrestrial microbes (all of the 20 would have been seen if the sample had been corrupted). Researchers discovered that the meteorite contains 18 percent more of the left-handed form of isovaline than the right-handed version, an indication that the preference for left-handed amino acids originated in outer space and developed when meteorites landed. Perhaps we owe the existence of proteins to interstellar material.

Proteolysis

Enzymes split protein into pieces known as *peptides*, which are smaller chains of amino acids. Eventually the peptides in cheese are degraded into the individual amino acids, which can then be converted into other products. The breakdown of proteins into smaller molecules is called *proteolysis* (pro-tee-OLL-lih-sis) and is the most important flavor-producing process in cheeses with harder textures. Amines and other nitrogen-containing compounds come from proteolysis of amino acids (Box 5.2), as do alcohols, aldehydes, ammonia, and compounds containing sulfur.

Fatty Acids

Fatty acids are formed from the breakdown of triglycerides. Fats, oils, triglycerides, cholesterol, and the fat-soluble vitamins (A, D, E, and K) are in a class of compounds are known as *lipids*, and the breakdown of these into fatty acids and other products is called *lipolysis* (lie-PAH-lih-sis). The butyric acid we saw in Box 1.1 is a fatty acid. Fatty acids not only have flavors and aromas of their own, but they are precursors of other aromatic compounds.

BOX 5.2
NITROGEN-CONTAINING COMPOUNDS

An assortment of compounds in cheese contain nitrogen. We have already looked at amino acids and ammonia. The next simplest nitrogen compounds in cheese are amines, which have carbon atoms included. Here is methylamine:

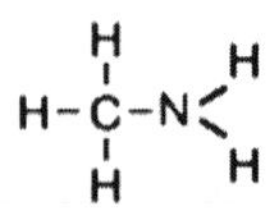

Next is ethylamine, with two carbon atoms:

NH_2

Then come propylamine (3 carbons) and butylamine (4 carbons). There are also dimethylamine, diethylamine, and dipropylamine. Diethylamine has this structure:

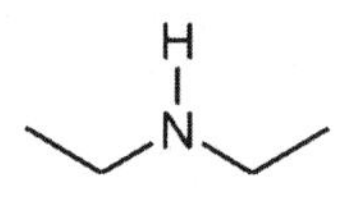

Triethylamine has a third ethyl group attached to the nitrogen atom, in place of the hydrogen atom. It has a fishy odor and can be detected at a concentration of half a part per billion.

Pyrazines contain rings with two nitrogen atoms and four carbon atoms. Most of the pyrazines in cheese impart earthy flavors. Here is 2-ethyl-3,5-dimethylpyrazine:

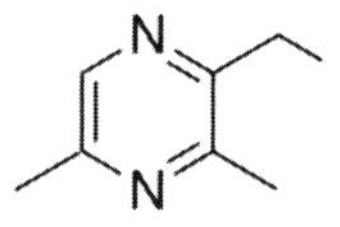

Positions here are numbered clockwise: the nitrogen at the top is at the number 1 position, ethyl (2 carbons) is at number 2, and the two methyl groups (1 carbon each) are at numbers 3 and 5. Other pyrazines in cheese have methyl and ethyl groups at different positions.

Indole and skatole are interesting compounds. At low concentrations, both have a flowery aroma and are used in perfume, but at higher levels they give human feces its characteristic odor. Yes, they are found in cheese. Here is skatole:

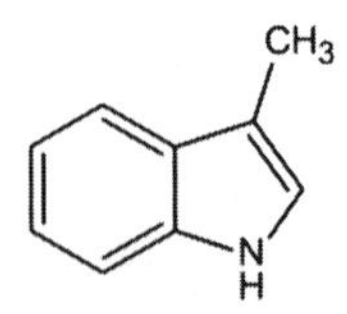

Indole has the same structure but without the CH_3. The word *indole* is a combination of *indigo*, the plant from which it was first isolated, and *oleum*, another name for fuming sulfuric acid, which was used to isolate it. *Skatole* comes from the Greek root *skato*, "excrement."

Here is a list of the fatty acids commonly found in cheese. The number in parentheses following the name is the number of carbon atoms in the molecule.

- Butyric acid (4), from the Latin *butyrum* (butter, since rancid butter contains it; "butane" is derived from this word). It has a cheesy, rancid odor that has also been described as "dirty sock."
- Caproic, caprylic, and capric acids (6, 8, and 10, respectively), from the Latin *caper* (goat, because of its odor). Their flavors are described as goatlike, rancid, waxy, and pungent.
- Lauric acid (12), from the Latin *laurus* (laurel, a main source of the compound). It is the first of the category called *long-chain fatty acids*, which do not greatly contribute to flavor, though their breakdown products do. (The smaller fatty acids are short-chain and medium-chain).
- Myristic acid (14), named after *Myristica fragrans* (the nutmeg tree, a major source).
- Palmitic acid (16), named after the palm tree. Coconut, palm, and palm kernel coconut oils contain a sizable amount of palmitic acid.
- Stearic acid (18), from the Greek *stear* (tallow).

These nine fatty acids are known as *saturated* because there are no double bonds between any of their carbons, allowing the carbons to be saturated with hydrogen atoms. The remaining three fatty acids are *unsaturated* because they do have double bonds, decreasing the number of hydrogen atoms in them.

- Oleic acid (18), from the Latin *oleum* (oil, since it is liquid at room temperature). Oleic acid has 18 carbon atoms like stearic acid, but contains a double bond in the middle that changes its properties. Its melting point is 56°, as opposed to 157° for stearic acid.
- Linoleic and linoleic acids (both 18), from the Greek *linon* (linen, since the flax plant contains it) + oleic. Linoleic acid has two double bonds and linolenic acid has three. The structures of stearic, oleic, linoleic, and linoleic acids are compared in Box 5.3.

You will notice that all of these fatty acids except propionic have an even number of carbon atoms. There are small amounts of odd-numbered fatty acids in milk and cheese such as the fifteen-carbon pentadecanoic acid. Living things manufacture fatty acids by adding two carbons at a time, which favors even-numbered carbon chains. Straight chains are also favored, although there are a few important branched-chain fatty acids in cheese (Box 5.4). Branched-chain fatty acids are also found in unexpected places (Box 5.5).

BOX 5.3
SATURATED AND UNSATURATED FATTY ACIDS

Here is a comparison between saturated and unsaturated fatty acids containing 18 carbons. When a nutrition label lists "saturated," "monounsaturated," and "polyunsaturated" fats, it refers to fats such as these. The position numbers shown are counted from the COOH.

Stearic acid has the full complement of 36 hydrogen atoms, making it a saturated fatty acid:

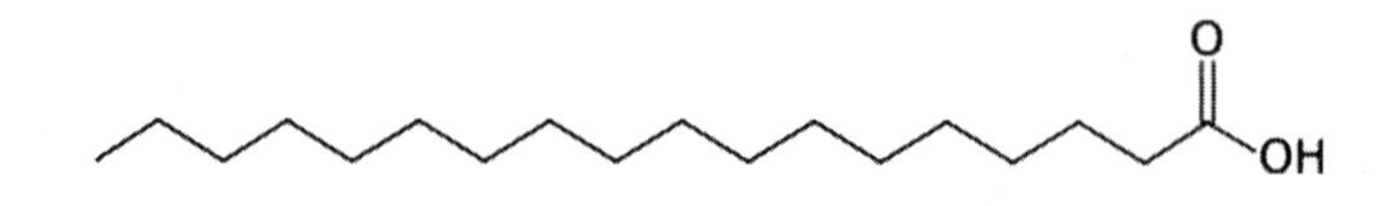

Oleic acid has a double bond after the ninth carbon atom and is sometimes called an omega-9 fatty acid. It is missing the two hydrogen atoms that would have been bonded to the ninth and tenth carbons. Oleic acid is a monounsaturated fatty acid since there is one double bond.

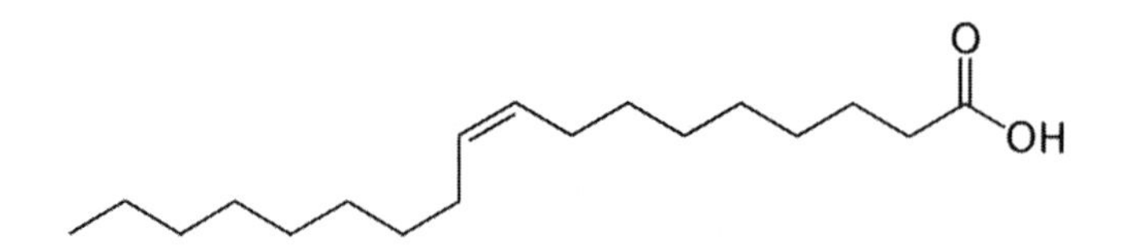

Linoleic acid has double bonds after the ninth and twelfth carbons, and is referred to as an omega-6 fatty acid. With the two double bonds, it is a polyunsaturated fatty acid.

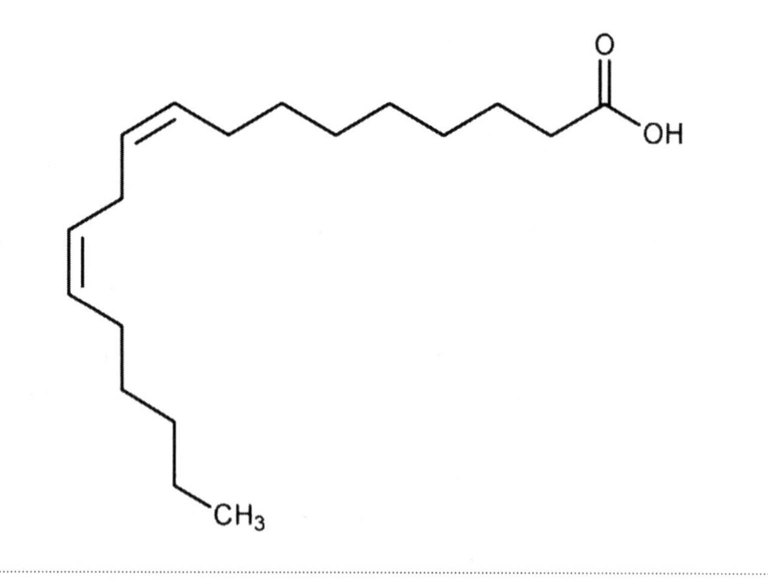

BOX 5.3 (*continued*)

Linolenic acid, an omega-3 fatty acid, is also polyunsaturated, with double bonds after the ninth, twelfth, and fifteenth carbons.

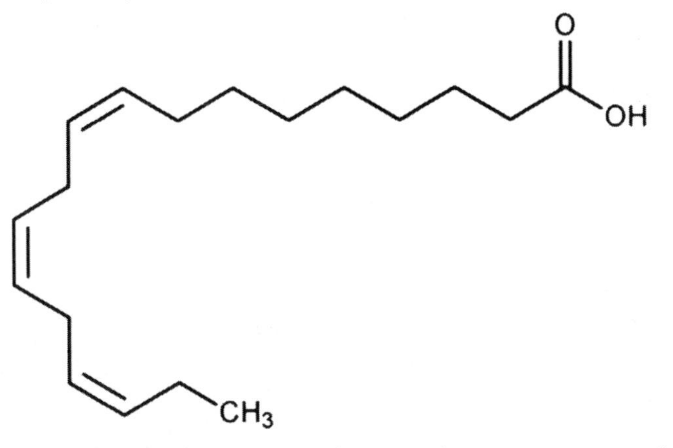

The hydrogen atoms around double bonds are in either a *cis* or a *trans* configuration. *Cis* is Latin for "on the same side"; all of the structures above have *cis* double bonds in their carbon chains. *Trans* means "on the other side," and the atoms around a *trans* double bond look like this:

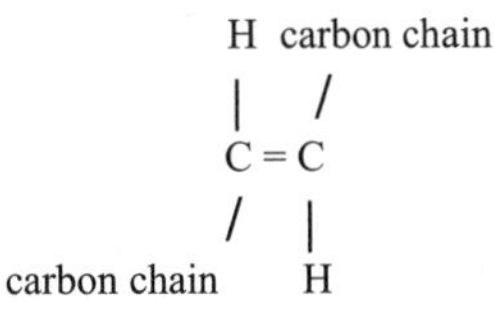

Oleic acid with a *trans* double bond is called elaidic acid, and the molecule is straight instead of bent. Elaidic acid is found in partially hydrogenated vegetable oil, which by far is the largest dietary source of *trans* fatty acid in humans. Vaccenic acid, with a *trans* double bond after the eleventh carbon, naturally occurs in milk. Elaidic acid has been implicated in decreasing "good cholesterol" (HDL, high-density lipoprotein) and increasing "bad cholesterol" (LDL, low-density lipoprotein); vaccenic acid has not.

Rumenic acid has a *cis* double bond after the ninth carbon atom and a *trans* double bond after the twelfth, making it both a *cis* and a *trans* fatty acid. Over 90% of the CLA in milk is rumenic acid (also called bovinic acid), with this structure:

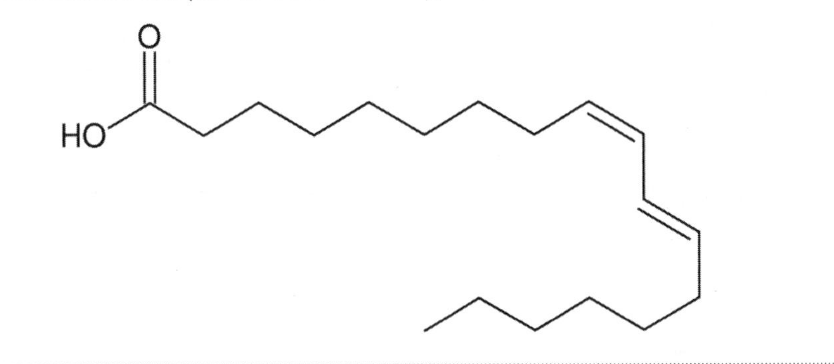

(*continued*)

BOX 5.3 (*continued*)

A molecule such as this where double bonds alternate with single bonds is referred to as *conjugated.* Rumenic and similar *cis, trans* fatty acids are called *conjugated linoleic acids* (CLA), and are important since they seem to have anti-cancer properties while reducing inflammation and risk for cardiovascular disease. The nutritional aspects of fatty acids will be covered in more detail in Chapter 14.

BOX 5.4
BRANCHED-CHAIN FATTY ACIDS AND HYDROXY ACIDS

Most of the fatty acids encountered in cheese are of the straight-chain variety as listed in the text, but there are a few branched-chain ones.

Isobutyric acid, also called 2-methylpropanoic acid, has a sweaty, rancid odor, and its structure looks like this:

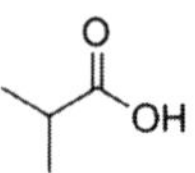

Isovaleric acid, or 3-methylbutanoic acid, has an aroma described as sweaty or cheese-like. Here is its structure:

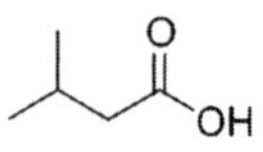

Two branched-chain fatty acids have pronounced goat flavors, and are found in goats' milk and sheep's milk but not in cows' milk:

4-methyloctanoic acid

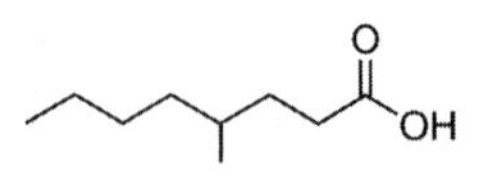

and 4-ethyloctanoic acid

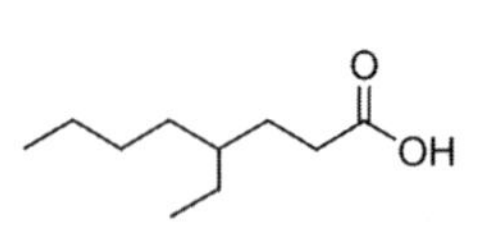

Small amounts of hydroxy acids are also found in milk. These are fatty acids with an OH group attached at the fourth or fifth carbon atom, counting from the COOH end. 4-Hydroxy and 5-hydroxy fatty acids rearrange into lactones, which we will encounter in Chapter 11.

The fatty acid content in milk varies according to species. See the bar graph (Figure 5.3) showing the percentages of fatty acids in the milkfat from cows, goats, and water buffalo.

This fatty acid profile shows that sheep's milk and goat's milk contain significantly more caprylic, capric, and lauric acids than does cow's milk. These fatty

BOX 5.5
STINKY PLANTS

Many plants give off pleasant (to us) aromas to attract certain pollinators. Others let loose with odors that may charitably be called "disagreeable." Notable among these are some plants in the Araceae (or Arum) family, including species of Amorphophallus (Greek for "misshapen penis," referring to the spike projecting from it). The smells coming from these tubers are described as "dung," "nauseating gas," "rancid, rotting meat"—and strong cheese. *Amorphophallus elatus* in particular produces a powerful cheesy odor, and the only compound responsible is isohexanoic acid, also called isocaproic acid:

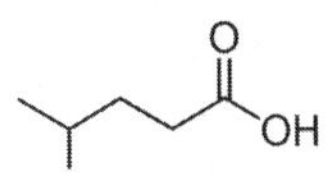

Isohexanoic acid has a pungent cheese aroma. Isovaleric acid, which has one less carbon atom, is found in several species of foul-smelling flowers. Flies are the pollinators of these plants, so we may assume they, like some people, are attracted to the odor of cheese. A couple of other Amorphophallus species generate cheesy fragrances, which have attributed to combinations of acetic acid, dimethyl disulfide (which we will meet in Box 7.1), ethyl acetate (Box 8.3), and 2-heptanone (Box 9.2).

Arum plants.
Source: Author's collection.

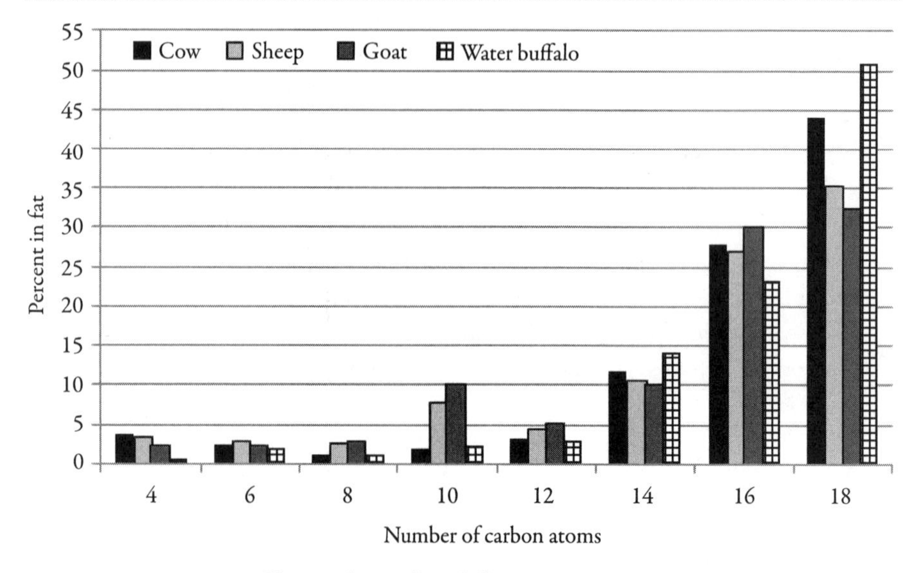

FIGURE 5.3 Percentages of fatty acids in milk in different mammals.

acids, along with butyric and caproic, are responsible for a good deal of cheese flavor. Other things being equal, a cheese made from sheep's or goat's milk would have a stronger flavor than one made from cow's milk. The fatty acids with 14–18 carbons are more plentiful in milk than the others, but are also much less flavorful.

The plants that animals eat influence the fatty acids in cheese, as do manufacturing procedures. Mozzarella is a bland cheese with less than 400 ppm of fatty acids, whereas its cousin Provolone, which has added enzyme and is aged, contains more than 2100 ppm. Parmesan has around 5000 ppm fatty acids.

Lipolysis

Certain enzymes cause fatty acids break down into other compounds, including alcohols, aldehydes, esters, ketones, and lactones. An enzyme in milk called *lipoprotein lipase* causes lipolysis in cheeses made from homogenized milk, since homogenization removes the membrane protecting milk fat globules (Box 2.2). Lipoprotein lipase is found in raw milk cheeses since pasteurization mostly inactivates it. Cheeses such as Feta, Pecorino Romano, and Provolone are made with a paste of rennet containing lipase, resulting in a good deal of lipolysis. The starter and adjunct cultures mentioned in Chapter 2 also break down fatty acids, with *Penicillum roqueforti* being the most effective. The fatty acid level in Roquefort cheese has been measured at over 26,000 ppm, more than any other variety.

Salt

Proteolysis and lipolysis are affected by NaCl concentration, so this is a good place to talk about salt. NaCl is added to cheese to inhibit or stop microbial growth and enzymatic activity while affecting the moisture content, the flavor, and state of the casein. The dehydration caused by diffusion of sodium chloride is the reason that it has been used since ancient times as a preservative in many foods, including cheese. Microbial functions are halted or greatly lessened when salt draws some of the water out of a cell, and enzymatic activity is inhibited as well.

You will remember from Chapter 2 that there are three ways of salting cheese: by dry-salting the curd, rubbing NaCl on the surface, and brining. Sodium chloride diffuses into cheese and causes moisture to diffuse out, balancing the NaCl and H_2O concentrations on the surface and just below it. When dry-salting or rubbing, NaCl dissolves in the moisture on the exterior of the cheese and starts to penetrate into the curd, sending whey out of the curd and creating a brine solution on the surface. The remaining salt crystals then dissolve and also start to diffuse into the cheese. Surface rubbing is conducted over several days, but a piece of curd with an irregular surface will allow salt absorption from many directions at once, so salt penetrates through it in about 20 minutes. In brine salting, the center of the cheese block initially contains no NaCl, and it takes up to five days for the salt to penetrate. The speed and amount of salt uptake increases with temperature, up to a point: above 90° melting fat on the surface blocks some of the NaCl from entering. Shape also affects salt diffusion; we saw in Chapter 2 that a cheese with corners allows NaCl to enter from more directions than a curved piece.

The amount of salt added to cheese depends on the microbes in it and the extent of growth required. For example, Emmental and Gruyère contain only 1.2 percent NaCl whereas Domiati and some blue cheeses have at least 4.5 percent. The variation is due in part to the levels of proteolysis and lipolysis desired. Strains of Lactococcus and Penicillium are stimulated at 1 percent NaCl and inhibited at 5 percent; and plasmin, the principal enzyme in milk that survives pasteurization, is stimulated at 2 percent NaCl and still displays some activity at 8 percent. Therefore, the amount of sodium chloride in cheese has to be adjusted to create a product with the right balance of flavors.

Salt directly affects flavor, and is the only basic taste in cheese that amino acids do not contribute toward. NaCl also seems to increase the amount of water held by the casein in the cheese, which has been found to help with cooking properties of Mozzarella. When Mozzarella has too little salt, it does not stretch enough when heated, and the oil is not liberated as much. We will explore Mozzarella in the next chapter.

To recognize good cheese: Not at all white like Helen, nor weeping, like Magdalene. Not Argus, but completely blind, and heavy, like a buffalo.
—ANONYMOUS, *Le Ménagier de Paris*, 1393

6

STRETCHED CURD CHEESES, ALCOHOLS, AND MELTING

IN *LE MÉNAGIER*, a husband addresses his younger wife about marriage and housekeeping, and makes sure she selects cheese that has some color (Helen of Troy was famed for her gleaming white skin), that does not have whey coming off of the surface (Magdalene bathed Jesus' feet with her tears), and does not have eyes (Argus was a mythical giant with multiple eyes). The buffalo probably referred to the wisent, or bison d'europe, but it is not a stretch to associate the reference to the water buffalo. Water buffalo milk is traditionally used in Italy for manufacturing Mozzarella, the most popular stretched-curd cheese. This category of cheese is also called *pasta filata* ("spun paste") or pulled curd because of the kneading required after the whey is drained. The stretching step enhances melting, which makes Mozzarella the top choice for topping a pizza.

Naples is credited as the place of origin of Mozzarella, perhaps as far back as the early fifteenth century. According to legend, the first Mozzarella came about after some curd was accidentally dropped in a bucket of hot water. Provolone originated in southern Italy in the nineteenth century and Caciocavallo-type cheese was mentioned by Hippocrates in 500 BC. Scamorza is a drier version of Mozzarella, though the compositions of the two can legally be the same in the United States.

One does not ordinarily associate alcohol with Mozzarella or Provolone, but one alcohol in particular, 1-octen-3-ol, is responsible for a significant portion of the flavor. And ethanol, which is created in alcoholic beverages by fermentation of sugar, arises in these and other cheeses through metabolism of lactose.

BOX 6.1
ALCOHOLS

Alcohols have an *OH group*, an oxygen atom bonded to a hydrogen atom, somewhere in the carbon chain. Ethanol is CH_3CH_2OH, and propanol, which has a sweet odor in cheese, is $CH_3CH_2CH_2CH_2CH_2OH$. These are "straight-chain" alcohols, with no forks or branches. Pentanol (fruity or alcoholic odor) has five carbon atoms, hexanol (slightly fruity or green) has six carbons, and heptanol (unripe fruit) has seven. Other alcohols responsible for cheese flavor have branched chains. Here are three major ones:

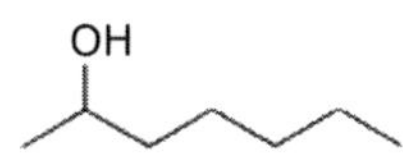

2-Heptanol has the OH group at the second carbon atom. The flavor has been described as earthy, green, fruity, and sweet.

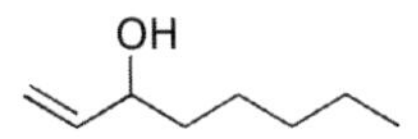

1-Octen-3-ol has a double bond after the first carbon atom and an OH group at the third one. It is also called *mushroom alcohol* since it is produced by, and has an unmistakable odor of, edible mushrooms.

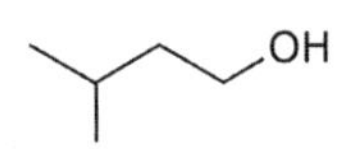

One of the most common alcohols in cheese, 3-methylbutanol has a methyl group at the third carbon atom and imparts a fruity or alcoholic flavor.

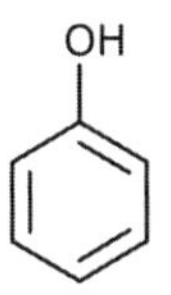

In addition, some alcohols contain a phenyl group—a circle of six carbon atoms with what amounts to 1½ bonds between them. It is depicted as a hexagon with three double bonds inside. These rings are common in nature, and molecules containing them are called *aromatic compounds* because they often have aromas. The simplest alcohol with a phenyl group is phenol, shown here. It is used in disinfectants, and its odor is commonly described as medicinal, though in cheese it may carry a livestock odor.

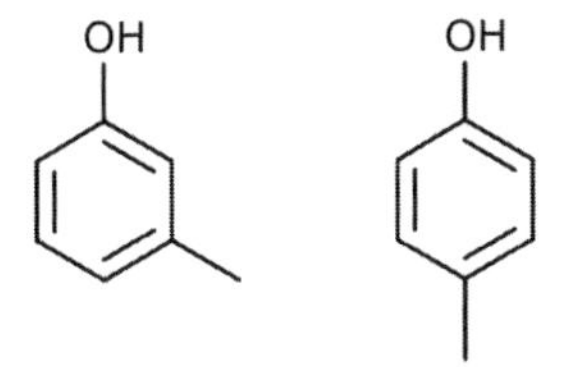

(*continued*)

BOX 6.1 (*continued*)

A pair of compounds, *meta*-cresol (with CH_3 on the side of the ring) and *para*-cresol (CH_3 opposite the OH), have aromas similar to phenol's. When the concentration is too high, cresols impart a "cowy" or "barny" odor.

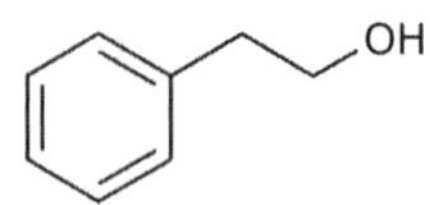

2-Phenylethanol has a phenyl ring attached to ethanol. This molecule smells like roses.

Mozzarella

In America, Mozzarella and Scamorza are made from cow's milk. Milk is cooled to 90°, starter is added, and the pH starts to decrease in 15–30 minutes. After rennet is added, the curd coagulates in 20–30 minutes and is then cut with quarter-inch knives. After 15 minutes, the temperature is raised to 106°, and curd is cooked until the pH reaches 6.1. The curd is sliced into slabs 6 inches deep and 18 inches wide along the sides of the vat. The slabs are cut in half and flipped every 15 minutes until the pH reaches 5.4. Originally, pieces of the curd were broken off by hand (*mozzare* is Italian for "cut off" and *scamozzare* means "to behead") and stretched multidirectionally under hot water. In mechanized production, the curd is chopped into small pieces and fed into a stretching machine that heats it to 122° and kneads it like dough. The starter generates enough lactic acid to dissolve CCP from the curd, allowing the curd to stretch at pH 5.25. The cheese is cut into the desired shape and floated in brine before packaging. When fashioned into tubes, it is marketed as string cheese. The Mexican version, Oaxaca (wah-HA-kuh, named after the Mexican state), is braided into balls. Most Mozzarella cheeses are vacuum-packaged, but some, especially those made from water buffalo milk, are submerged in cups of whey to maintain freshness (shown in Figure 6.1). These fresh Mozzarella cheeses are usually consumed within a week or two. Moisture is not tightly bound in Mozzarella immediately after manufacture, so packaged cheese is aged for a few days. Mozzarella must be held that long if it is to be shredded, for otherwise the wet shreds will stick to the equipment and each other.

It may surprise you to learn that Mozzarella, not Cheddar, is now the most popular cheese in the United States (Box 6.2). Much of the Mozzarella manufactured is low moisture (45–52 percent moisture) and part skim (30–45 percent fat in dried cheese) because this type does not release much whey or oil when baked on a pizza. Pizza is the most popular school lunch, and the National School Lunch Program is

FIGURE 6.1 Water buffalo Mozzarella in a cup of whey.
Source: Author's collection.

trying to combat childhood obesity by purchasing only "lite" (less than 10.8 percent fat) and part skim Mozzarella for schools.

Provolone and Caciocavallo

Provola (from "globe-shaped") is also a stretched curd cheese, but the better-known American variety is Provolone, which simply means "large Provola." Rennet is added immediately after the starter, and the curd is cooked at 118°–126°. The cheese may undergo a smoking process. Provolone is traditionally aged while hanging from a string, with one cheese tied to each end, and the surface is frequently hand-rubbed with an oily cloth soaked in brine. The lengthy ripening of this variety provides it with a more pronounced flavor than Mozzarella's.

Caciocavallo means "cheese on horseback" because this variety is traditionally aged by dangling on either end of a horizontal stick or rod, resembling saddlebags on a horse. It is fashioned into a teardrop or gourd shape for ease of hanging. Caciocavallo is made from cow's milk using a procedure like those of Mozzarella and Provolone, and is aged from three months to two years. The similar Kashkaval is made in southeastern Europe from sheep's milk.

BOX 6.2
PRODUCTION

The production of cheese in the United States has increased faster than the population, as you can see in the chart below. In the 1950s and 1960s, Americans were consuming less than 10 pounds of cheese per capita, so there were more than ten times as many people as pounds of cheese produced. In 1970, there were 2,201,430,000 pounds of cheese made for 205,052,000 people, or 10.7 pounds per person. Cheese production and consumption has increased dramatically since then: in 2011, manufacturers produced 10.609 billion pounds for 311.6 million people, or 34 pounds per capita. About 95% of this cheese is being eaten by American consumers. The United States exported 494 million pounds in 2011 and imported 313 million pounds.

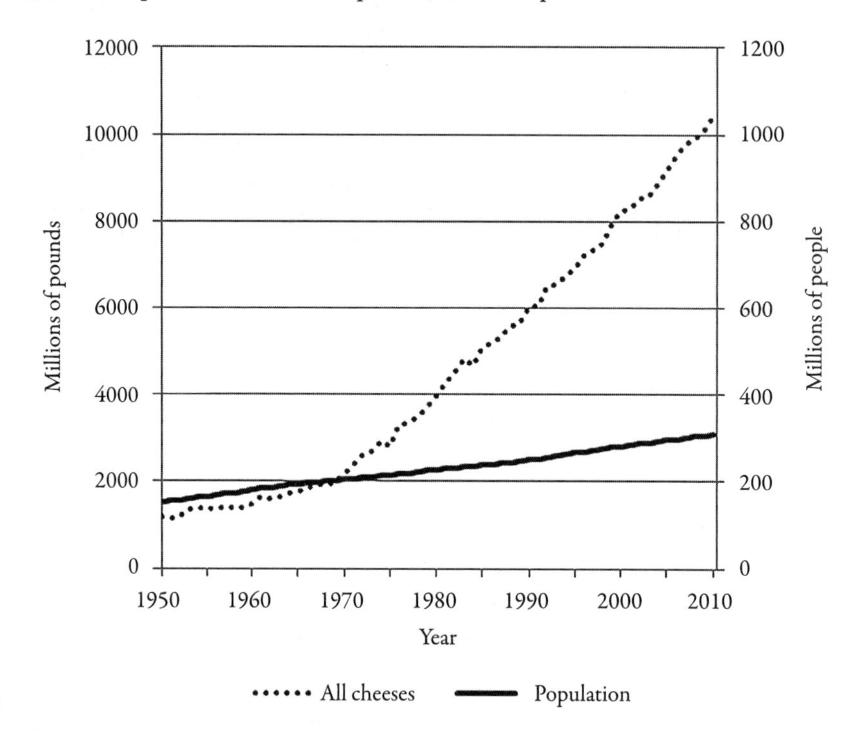

Cheese production in the United States by year.

When you ask Americans to name a variety of cheese, many will choose Cheddar; and if you asked them to draw a picture of cheese, most will depict Swiss. But the most popular variety in the U.S. for the past few years has been Mozzarella, mainly due to the increasing popularity of pizza and string cheese. The chart below shows that production of Mozzarella first surpassed that of Cheddar in 2001 and has stayed ahead every year since 2006. Cottage cheese, including creamed, dry curd, and lowfat, ranks third. The USDA reports Colby and Monterey Jack together, and they combine for fourth place. Cream and Neufchâtel are fifth, and Swiss is sixth.

BOX 6.2 (*continued*)

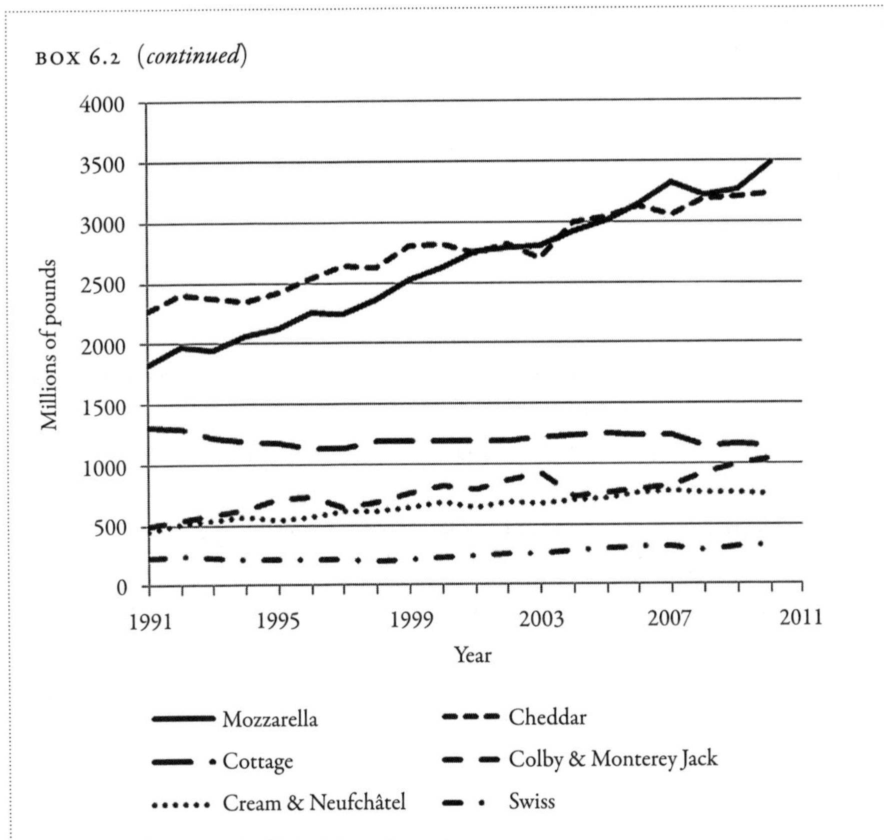

Cheese production in the United States by variety.

The top cheesemaking state is Wisconsin, with 2.637 billion pounds in 2011, followed by California with 2.245 billion pounds. Idaho and New Mexico have built large cheesemaking industries, and check in at third and fourth place with 841 and 744 million pounds, respectively. New York (741 million pounds), Minnesota (603 million), and Pennsylvania (412 million) are the other states producing more than a quarter billion pounds of cheese a year.

Alcohols

Alcohols in cheese arise from metabolism of lactose, amino acids, and fatty acids. A common alcohol in cheese is 1-octen-3-ol; this and other flavor compounds containing eight carbon atoms originate from linoleic and linolenic acids (Box 6.1). Its mushroom odor has been identified as important in Brie and Camembert (where it is a product of Penicillium metabolism) as well as Grana, Gorgonzola, and Mozzarella. 1-Octen-3-ol can be detected at a concentration of 10 ppb. The flavor of 1-octen-3-ol is enhanced by a similar compound, the ketone 1-octen-3-one. 2-Phenylethanol is a product of metabolism of phenylalanine and is present in Camembert almost immediately due to action of the yeast.

One study showed that water buffalo milk contains 55 ppb of volatile alcohols, about as much as milk from cows, goats, and sheep combined. Water buffalo Mozzarella therefore has higher levels of alcohols than cow's milk Mozzarella. Although alcohols are not the primary class of flavor compounds in milk (esters are), they still play an important role in cheese.

Melting

The stretching step in making stretched-curd cheeses aligns the protein matrix into parallel fibers, with the fat droplets and whey filling in the spaces between them. The separation of fibers decreases the number of connections between proteins, making the cheese easier to melt. The casein itself does not actually melt, but the increased movement of the fibers makes it behave in a more liquid-like manner. Fat droplets also decrease the number of connections by interrupting the continuity of the matrix, so melting improves as the fat content goes up. The liquefying fat allows the cheese to flow since the matrix no longer has solid fat to help support it.

The pH and amount of colloidal CCP have a bearing on the melting properties of any cheese variety. As the pH drops, more CCP goes from the casein into the whey, which means that more of the glue holding casein micelles together disappears, enabling the cheese to melt more readily. Cheese will not melt above pH 6.0, and it melts best around pH 5.0–5.4. It will still melt below pH 4.8, but the caseins start to reorganize into aggregates, preventing flow.

An increase in the ability of the casein to hold water enhances melting. Water-binding improves with age, as some of the caseins swell up and become soluble in water. The NaCl content plays an important role in this process. Melting also increases with proteolysis because the gradual breakdown of the protein structure allows fat droplets to coalesce. Figure 6.2 shows magnified images of a Mozzarella cheese made by our USDA research group and stored at 40° for one and six weeks. The gray areas correspond to the protein matrix, and the black areas to fat droplets. The fat is in the form of discrete globules after one week, but aggregates into elongated shapes over the following five weeks. You can also see some of the *Streptococcus thermophilus* starter culture inside many of the droplets at one week. These bacteria, like all other starter cultures, die off, break open, and empty their contents over time, releasing enzymes that attack the protein and fat. Almost no bacteria are visible at six weeks, though their enzymes have clearly had an effect.

Proteolysis was the key for the development of lite Mozzarella in our laboratory in the 1990s. Prior to then, a lack of melting and stretching was a key problem with low-fat versions of Mozzarella. Manufacturers had simply used low-fat milk, which increased the amount of protein in the cheese and decreased its ability to flow

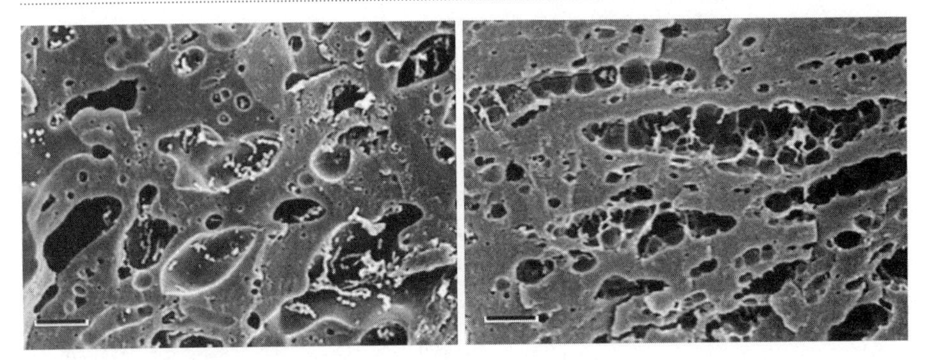

FIGURE 6.2 Scanning electron microscope images of Mozzarella cheese after refrigerated storage for one week (left) and six weeks (right). The black lines at lower left correspond to 10 micrometers. *Source:* Author's collection.

when heated. By lowering the temperatures at which the cheese is cooked and stretched, less of the starter culture and rennet was inactivated, allowing for more proteolysis and therefore more melting and stretching. Another problem with low-fat Mozzarella was a lack of flavor—most of the flavor compounds in cheese are found in the fat rather than the whey, so a reduction in fat meant a reduction in flavor. We added an extra Lactobacillus culture that imparts a strong cheese flavor, which solved the problem.

Browning

Mozzarella and other cheeses turn brown when heated sufficiently because of Maillard ("MY-yard") reactions, a series of chemical reactions between amino acids and sugar, and not involving enzymes or microorganisms. The toasting of bread, browning of steak, aroma of coffee, and flavor of chocolate are all due to Maillard reactions. A slew of flavorful products is also generated (Box 6.3). Other reactions create molecules that impart yellow and brown colors to the cheese, which explains why cheese develops a deeper color during long storage periods.

Free Oil

Free oil is the puddle of liquid fat that you may find atop a freshly baked pizza. The protein matrix contracts as cheese is warmed, pushing out whey and creating gaps that allow oil to flow out also. The oil does not flow as readily in cheeses that do not have their curd stretched since the fat is distributed evenly throughout the matrix and not between fibers. The oil can be liberated from these varieties by heating higher and faster. The extent of free oil formation increases with fat content (because

BOX 6.3
MAILLARD REACTIONS

This important series of reactions in food was first described by French physician Louis Camille Maillard in 1910. The first step is a reaction of a carbohydrate and an amino acid; in cheese, the carbohydrate is lactose, and the amino acid is primarily lysine. One of the rings in lactose opens up, and a C=O group (an arrow points to it in the diagram) is formed:

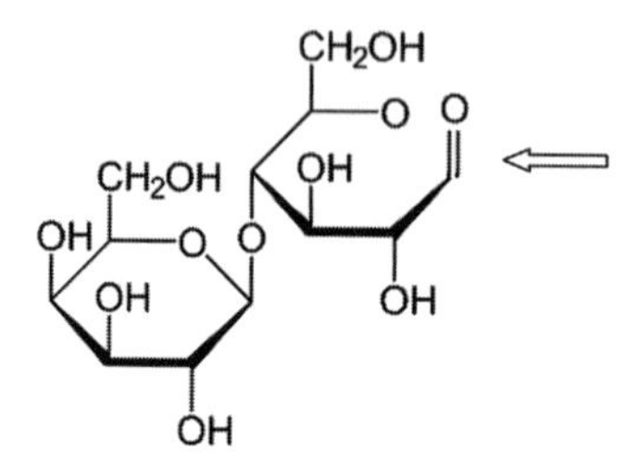

The C=O reacts with an NH_2 group usually on the end of the lysine:

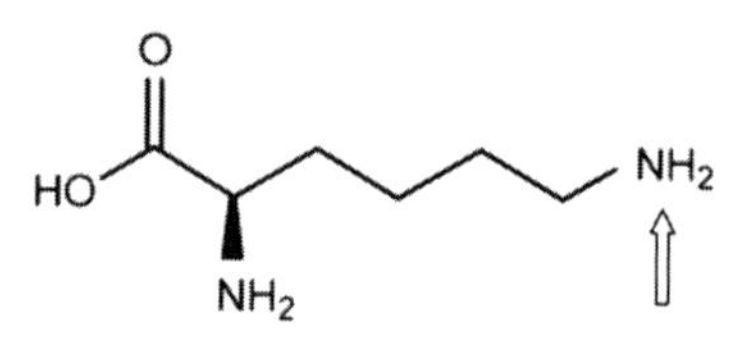

An unstable intermediate compound forms, containing both the amino acid and the lactose, and it then breaks down into a multitude of molecules containing carbon, hydrogen, oxygen, and nitrogen, often in rings. The nitrogen-containing compounds include pyrazines (Box 5.2), and the ones with oxygen include furans (we will see these in Chapter 12).

the protein matrix can't hold all of the melting fat) and with age (as proteolysis takes place).

Stretching

Cheese can stretch when there is a continuous and intact casein matrix, but it loses that ability as proteolysis becomes more extensive. When you pull on two ends of a piece of warm Mozzarella, casein molecules break apart, slide past each other, and connect with other casein molecules, repeating the process many times. Stretchability is an important reason Mozzarella is used on pizza; consumers reject pies with non-stretching cheese. Cheese can stretch between pH 5.8 and 4.8, with 5.2–5.3

being the optimum. At that pH level, around 75 percent of the CCP has been liberated from the casein, which is still in a continuous network, resulting in a cohesive and stretchable curd. Proteolysis breaks the continuity of the network, reducing stretchability.

Stretched-curd cheeses have strong and elastic structures owing to their manufacturing techniques. The same cannot be said for Brie, Camembert, and other surface mold cheese, which can become runny after a while. We will learn about them in Chapter 7.

Sometimes I think I'm liquefying like an old Camembert.

—GUSTAVE FLAUBERT, two weeks before his death in 1880

7

SURFACE MOLD CHEESES, SULFUR COMPOUNDS, AND THE SENSES

THE TWO COMMON surface mold cheeses available in the United States, Brie and Camembert, will liquefy inside if allowed to sit around for too long. The bacteria, mold, and yeast added to these varieties are responsible for a rich eating experience, with characteristic appearances, aromas, flavors, and textures. The interplay of sensory inputs when eating cheeses such as these is the reason people enjoy eating them.

This class of cheese appears to have been created in the Middle Ages by peasant farmers, who kept only a cow or two and had to store milk for several days to make cheese. The older milk began to sour, not only forming a natural starter culture but also creating a good environment on the cheese surface for molds, which are more tolerant of acid than bacteria.

Camembert

Brie and Camembert are traditionally made from raw milk, although the large manufacturers in France apply heat treatment or filtration for safety. Camembert is named after a village in Normandy where legend claims that it was invented by Marie Harel in 1791. Though a statue of Harel stands in the town, the cheese appears to predate her by at least a century. Camembert is made with milk that is not allowed to exceed 100°. Lactic acid bacteria are allowed to grow in milk for up to a day at 50°, allowing the pH to decrease below 6.5. The temperature is then raised, rennet and *Penicillium camemberti* spores are added, and the curd forms. As the pH continues to drop, the CCP becomes soluble and dissolves in the whey, producing a fragile curd. The curd is not cut, but is hand-ladled into circular forms to produce cheeses 4½ inches in diameter, an inch thick, and weighing half a pound. The curds drain for 4–5 hours, the forms are flipped over, and the curds drain overnight without being pressed. The cheeses are removed from the forms, dry-salted, and placed on shelves in a low-humidity room at 54° for up to a week to dry the surface and allow for mold to grow. The pH starts at 4.6–4.7. *Kluyveromyces lactis* yeast, *Geotrichum candidum* mold, and *P. camemberti* bacteria metabolize the lactose and break down some amino acids to ammonia, with the net effect of decreasing the acidity. When the pH is above 5.8, Staphylococcus or Coryneform bacteria, which can adapt to the high salt content, grow further and ripen the cheese. *P. camemberti* becomes visible as a white, felt-like growth on the surface after a week. As mentioned in Chapter 2, the CCP becomes insoluble as the pH increases and is drawn to the surface, causing the center of the cheese to soften. The product is wrapped in wax paper, transferred to wooden or cardboard boxes, and stored at 45° for another week. Proteolysis and lipolysis are quite extensive, generating a slew of flavor compounds. If held too long, the breakdown of the protein matrix will result in a runny cheese.

Brie

The Brie region east of Paris is the birthplace of Brie cheese, which may date back as far as the eighth century AD. It is made in a similar manner to Camembert with some differences in the details. Wheels of Brie are larger, with diameters ranging from 9 to 15 inches, thickness of 1 inch, and weighing 1–6 pounds. The larger size will cause Brie to age a little slower and produce a slightly blander flavor. Coulommiers (KOO-lohm-yay, named after the town where it was developed) is similar to Brie but is made in different sizes and has no standard manufacturing procedure. Consumers often trim away the rinds on Brie and Camembert, but they are edible.

A variation on Brie is Brillat-Savarin, named after French gourmet Jean Anthelme Brillat-Savarin (appropriately born in Belley, France, in 1755). This is a triple crème cheese made from milk with enough added cream so that the product contains about 40 percent fat. It is aged only a week or two.

Dusting with Ash

Morbier (named after a French village) was traditionally made by taking soot from the base of the kettle and layering over the curds from the morning milk to protect it from insects and drying out. Curd from the evening milk would then be placed on top, resulting in a dark horizontal stripe through the cheese. Nowadays, vegetable ash or food coloring is added after the wheel is cut horizontally. Humboldt Fog, originating in Humboldt County, California, emulates Morbier by filling the forms halfway with curd, sprinkling with vegetable ash, and adding more curd. Humboldt Fog is made with batch-pasteurized (30 minutes at 145°) goat's milk, and Morbier with raw cow's milk.

Sulfur Compounds

Surface mold cheeses contain aldehydes, fatty acids, ketones, and lactones from lipolysis; amines and sulfur compounds from proteolysis; and alcohols and esters from both processes. 1-Octen-3-ol and the similar 1-octen-3-one provide mushroom aroma in a higher concentration than in stretched-curd cheeses. But unlike Mozzarella and Provolone, sulfur-containing compounds are prominent in Brie and Camembert (and the smear-ripened cheeses in the next chapter), and are essential components of the garlic/cabbage flavor.

Methionine and cysteine are the sulfur-containing amino acids in milk proteins. The level of cysteine in cheese is low: α_{s1}-, β-and κ-caseins do not contain it, although α_{s2}-casein and whey proteins do. There's plenty of methionine, however, and *Geotrichum candidum* and *Penicillium camemberti* degrade it to methional and methanethiol (Box 7.1). Both are important odorants in Brie and Camembert and precursors for other key compounds. At the concentrations found in cheese, methional has a cooked-cabbage aroma and methanethiol has a boiled-potato odor. They can be detected at a level of tenths of a part per billion. At higher concentrations, methional is quite pungent and methanethiol can clear a building. Methanethiol, also known as methyl mercaptan, is added to natural gas to alert people of a leak. When I was an undergraduate in college, a classmate synthesized some in a fume hood, following correct procedures. But the odor went through the exhaust on the roof and wafted downwind to the air intake of a 14-story office building nearby, which was evacuated for fear that leaking gas would blow up the place.

The Senses

Consumer acceptability of food depends on the five senses. The appearance, the aroma, the flavor, the sound made when bitten and chewed, and the texture in the mouth must combine for a pleasing experience.

Appearance is important when eating, since people look for specific colors and patterns in their food and are repelled by food that doesn't look right. In a study

BOX 7.1
SULFUR-CONTAINING COMPOUNDS

Of the compounds in cheese that contain at least one sulfur atom, the two most important are methanethiol and methional. A *thiol* contains a sulfur atom bonded to a hydrogen atom and a carbon atom; methanethiol, CH_3SH, is the simplest. Thiols are also found in beer, wine, and some foods. A notable thiol in cheese is S-methyl thioacetate, which has an aroma described as cooked cauliflower, cabbage, cheesy, and crab:

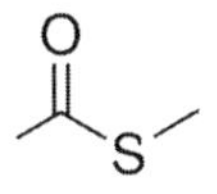

If you add another carbon to make S-methyl thiopropionate, you get a compound that smells just like Camembert:

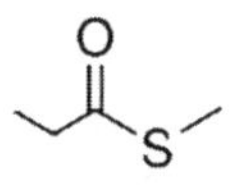

A thiol that was recently discovered in Camembert and Munster is ethyl-3-mercapto-propionate, which is also found in wine and Concord grapes. It has a fruity, grapey, rhubarb aroma:

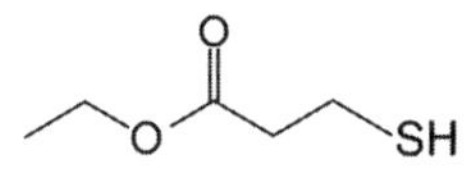

Methional is an aldehyde, and its structure is:

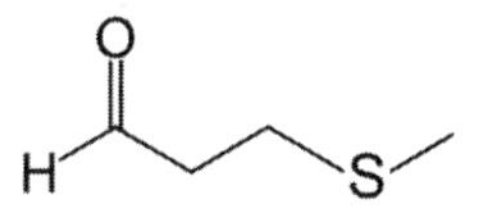

A compound that has both sulfur and nitrogen is 2-acetylthiazoline. It has a popcorn aroma:

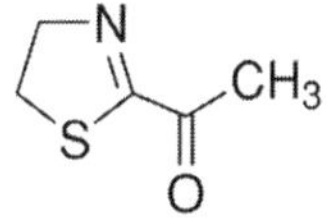

Dimethyl sulfone has a sulfurous or burnt odor:

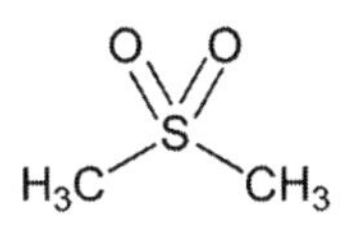

The other major sulfur compounds in cheese are sulfides. Hydrogen sulfide is H_2S and has a rotten-egg odor. It is thought to come from degradation of cysteine.

Dimethyl sulfide has this structure:

$$H_3C{-}S{-}CH_3$$

Dimethyl disulfide has two consecutive sulfur atoms, and dimethyl trisulfide has three. All three compounds impart sulfurous, cabbage, and onion aromas.

of Cheddar cheeses, panelists rated appearance as the leading factor, with color being most important. The exact coloration was a dominant factor. With surface-mold cheeses, ripeness is signaled by the shift from white to creamy yellow in the interior.

Aroma originates in the olfactory receptors, a class of proteins found in neurons (nerve cells) in the nasal cavity. Humans have 347 of these receptors. When an odor molecule binds to a receptor protein, a signal is sent to the inside of the cell, which relays it to the brain for interpretation. Each receptor can be activated by a large range of odorant molecules, and an odorant can bind to several receptors, so we are able to detect many thousands of specific smells. The type, size, shape, and concentration of molecules can all be distinguished. That is why methional and methanethiol change aroma as they become more diluted, and why the octanol in Box 3.2 is described as "orange- and rose-like" whereas octanoic acid is characterized as "rancid and sweaty." Scientists are not certain how our receptors are able to detect odors, but do know that no two odors are perceived to be identical. A substance cannot be smelled unless it is volatile, which is why compounds with long carbon chains, such as stearic acid, have little aroma. Aroma release is affected by the way a food is chewed, its texture (which may prevent diffusion of some odorants into the nose), anatomy, and other factors. People seem to prefer complex scents, such as those in cheese, to simpler ones. Aroma perception is enhanced by inhaling *orthonasally* (through the nose) and *retronasally* (through the mouth) to bring odorants to the receptors.

Flavor consists of three factors. The first is *olfactory*, described above. The second, *taste*, encompasses five sensations: bitter, salty, sour, sweet, and umami (Box 7.2). Taste buds, of which we have 2000–8000, are covered with taste receptors that send information to the brain. The final factor is *trigeminal*, referring to the trigeminal nerve, also called the fifth cranial nerve. This factor is responsible for facial sensations and contributes to flavor with sensations of heat, cooling (an example is menthol), pungency (hot peppers), tingling (carbonated drinks), and astringency (tannins in wine). The three factors together make up what we call *flavor*, with olfactory and taste factors being key for cheese.

Sound is not usually encountered when eating cheese, but sometimes you can find NaCl deposits on the outside of smear ripened cheese; calcium-containing crystals inside Camembert, Cheddar, and Roquefort; and crystalline clusters of tyrosine inside well-aged cheese, resulting in some crunchiness. When a brittle or rigid piece of food is bitten, the structure is ruptured and a cracking sound emanates. The fracture continues until the piece fragments, the bite is stopped, or the crack encounters a hole or something soft. Sound is sensed by the conduction of air to the ear, via soft tissues in mouth, and through conduction via the jawbone.

BOX 7.2
TASTE

The gustatory sense is composed of five basic tastes, and possibly a couple of others. Bitter, sweet, sour, and umami can arise when proteolysis has proceeded to the point where casein have broken down into amino acids. Despite what was earlier believed, receptors for taste are found throughout the tongue and are not confined to specific areas.

Bitterness is the most sensitive of the five tastes and is commonly encountered in coffee, citrus peel, and tonic water, which contains the notably bitter compound quinine. Many toxic plants contain bitter compounds, but rejection of foods for bitterness by animals and humans does not depend on whether response the food is harmful or harmless. Rather, it depends upon how many bitter and potentially toxic compounds are in their diet. Carnivores, who rarely encounter toxic plants, have a low bitterness threshold and a low tolerance for poisonous compounds, whereas grazers, who cannot limit their dietary choices as completely, have a high threshold and tolerance. A number of amino acids found in cheese have a bitter taste: arginine, isoleucine, leucine, methionine, phenylalanine, tryptophan, tyrosine, and valine. These are not nearly so strong as other flavor compounds, and they must be present at hundreds of parts per million to be detectable. Oleic acid in cheese appears to mask bitterness in cheese by bonding with bitter compounds.

Sweetness is not only related to sugar, but also to aldehydes, ketones, and other compounds containing the C=O group. These include the amino acids alanine, glycine, serine, and threonine. Proline is sweet and bitter. These can't be detected unless they are at a level of tenths of a percent.

Sourness is a measure of acidity. Citrus and some other fruits are naturally sour, and so is cheese when the pH is low. Aspartic acid and glutamic acid are amino acids with sour tastes. These amino acids can be detected at 30–50 ppm.

Saltiness is related to NaCl and KCl, potassium chloride. KCl is used as a salt substitute for people who are reducing their intake of sodium, but the pure compound is bitter and imparts a metallic taste to some people.

Umami, a savory impression imparted by monosodium glutamate (MSG), comes from the glutamate form of glutamic acid. It wasn't identified until 1908 and was not officially recognized as a basic taste until the 1980s. Umami is described as stimulating the oral cavity and the tongue, on which there is a slight furriness. Umami enhances the flavor of food, which is why some chefs add MSG to their dishes, and it is a noted feature of ripe tomatoes, soy sauce, and aged cheese.

Two other sensations that may be basic tastes are fattiness and calcium. A particular gene in mice and rats allows them to detect lipids, which may mean that humans also possess a fat receptor. Michael Tordorf of Monell Chemical Senses Center in Philadelphia reported at an ACS meeting in 2008 that mice have a calcium receptor that helps control their calcium intake, and that humans may have one also. When I asked him later what calcium tastes like, he replied that it was "calciummy."

Texture is considered by food scientists to be the sensory manifestation of the structural, mechanical, and surface properties of foods, and is detected through vision, hearing, and especially touch. Particle size, shape, and orientation lead to perceptions of crystalline, fibrous, grainy, and gritty characteristics. Mechanical properties include adhesiveness (sticky/gooey), cohesiveness (brittle/chewy/gummy), elasticity, hardness/softness, and thinness/viscosity. Levels of moisture (dry/wet) and fat (oily/greasy) also contribute. Humans obtain information about the feel of foods by using their tactile sense from inside the mouth along with nerve signals from jaw muscles and joints.

God's Feet

Camembert in particular has been featured in the fine arts. Brillat-Savarin said, "Camembert, poetry / Bouquet of our meals / What would become life / If you did not exist?" The Cole Porter song "You're the Top" contains the lyrics "You're the nimble tread / Of the feet of Fred Astaire / You're an O'Neill drama / You're Whistler's mama! / You're Camembert." In his autobiography, Salvador Dalí claimed that Camembert inspired the soft, melting pocket watches in his famed 1931 painting *The Persistence of Memory*: "I adore Camembert precisely because when it is ripe and beginning to run it resembles and assumes exactly the shape of my famous soft watches." The French poet Léon-Paul Fargue (1876–1947) once remarked, "Camembert, this cheese that smells of the feet of the good God." The aroma of Camembert, as well as Limburger and some other cheeses, is indeed reminiscent of the odor emanating from feet. The explanation is in the next chapter.

Splendid cheeses they were, ripe and mellow,
and with a two hundred horse-power scent
about them that might have been warranted
to carry three miles, and knock a man over at two
hundred yards.
—JEROME K. JEROME, *Three Men in a Boat*, 1889

8

SMEAR-RIPENED CHEESES, ESTERS, AND AROMA

WHILE CAMEMBERT IS revered as the epitome of food in poetry, art, and song, Limburger is held (and smelled) in a different light. In satire, cartoons, and comedy movie shorts, it is regarded as the quintessential malodorous food (Box 8.1). In reality, Limburger, Brick, and other smear-ripened cheeses have relatively bland flavors—if you can get past their interesting scents.

Smear-ripened cheeses are thought to have been created over 1,000 years ago in monasteries, where they served as an important protein source. Monasteries had large herds that provided enough milk to make cheese from a single milking. Unlike the milk from small farms, the fresh milk had no time to become sour, allowing acid-sensitive bacteria to readily grow on the cheese surface. Experimentation led to cheeses with meaty flavor notes, which was desirable for periods of religious fasting when meat could not be consumed. Smear-ripened cheeses tend to be made with large surface areas so that bacteria have plenty of space to work. They are typically brown, orange, or red due to the production of carotenoid pigments by microorganisms such as *Brevibacterium linens*, with the color being affected by the yeast. American Munster cheese is not smear-ripened as it is in Europe, and it is rubbed with annatto to impart the characteristic orange color.

BOX 8.1

LIMBURGER IN POPULAR CULTURE

In the mid-1800s, German and Belgian immigrants began to arrive in the United States with a taste for Limburger, which they enjoyed with raw onions and mustard in rye bread sandwiches accompanied with cold beer (to this day, aficionados claim this is the best way to eat it). By the 1880s, comedians began to associate Limburger with accented immigrants. Americans picked up on the idea that Limburger was a highly odiferous food that can be the butt of jokes, and have never let it go. In Mark Twain's 1882 tale "The Invalid's Story," the narrator shares a railroad car with another man, a Limburger cheese, and a pine box they thought contained a recently deceased man. The smell from "the deadly cheese" overpowers them, and the story concludes: "My health was permanently shattered; neither Bermuda nor any other land can ever bring it back to me. This is my last trip; I am on my way home to die." The first record ever sold by

Limburger cheese.
Source: Author's collection.

BOX 8.1 *(continued)*

the Victor Talking Machine Company was "Limburger Cheese" in 1901, a monologue by comedian Burt Sheppard. He relates that, as a boy, he bought some Limburger, placed pieces in his parents' pockets before they went to church, and watched as the congregation gradually fled. In the 1918 movie *Shoulder Arms*, Charlie Chaplin plays a World War I soldier who receives Limburger in a package from home, dons a gas mask, and lobs it into a trench full of German soldiers to force their surrender.

When Prohibition was enacted in 1920, workers could no longer patronize bars to eat their Limburger sandwiches, and the popularity of the cheese never recovered. It suffered a further blow in 1935 when newspapers revealed that Limburger was the favorite snack of Lindbergh baby kidnapper Bruno Richard Hauptmann. Over 6.5 million pounds of Limburger were produced in America in 1951, but it is now made in this country by only one company, in Monroe, Wisconsin, which turns out less than 650,000 pounds a year. Brick cheese, which gives off a less powerful aroma and has been called "the married man's Limburger," is manufactured at ten times that rate.

Limburger and Brick

Limburger was developed in the early nineteenth century in the Duchy of Limburg, now part of Belgium, Germany, and the Netherlands. The milk is cooled to 90°, and starter and rennet are added. After 30 minutes, the curd is cut with 3/8-inch knives, and allowed to rest 10 minutes. It is then held at 98° for 30 minutes. After brining or dry salting, the curd is placed in forms and undergoes a low-temperature "sweat" for 1–4 hours by being placed in a 68°–72° room at 95–98 percent relative humidity. A culture of *B. linens* and other microbes is applied to the surface by smearing with a brush or by soaking with a high-pressure spray. The cheese is stored at 55°–59° and 95–98 percent relative humidity for two weeks, during which it is washed with brine several times. The humid conditions allow the bacteria to proliferate, and the brine washing controls their spread and inhibits mold growth. Moreover, yeast in the air settles on the surface and metabolizes lactic acid, raising the pH. The cheese is wrapped in aluminum foil and stored under refrigeration for up to two months. Liederkrantz, which we met in Box 2.7, is a milder version of Limburger. Production was suspended from 1985 to 2010, but it is now being made by a company in Richfield, Wisconsin.

Brick cheese, invented in Wisconsin in 1877, is brined for 24 hours and not subjected to the sweat step. It is traditionally pressed using actual bricks (its name was also influenced by its shape) and is smeared six times over six weeks. The minimum pH of Brick and Limburger is a relatively high 7.0.

Reblochon

Reblochon is from an old French word for "re-pinch the udder" and dates to the 1200s. During the following century, farmers in the Savoy Alps would partially milk their cows in the presence of the tax-collecting landowner, pay the levy based on the milk produced, and later complete the milking after the taxman left. The second batch was richer and used for this cheese, which was made solely for family consumption, until the tax was lifted after the French Revolution. The cheese is made from unpasteurized milk from cows that graze on pasture in the Savoy during the warm months and eat hay cut from those pastures in the cold months. The milk is heated to 99°, starter and rennet are added, and the curd is cut into rice-sized pieces. Cheesecloth is stretched over a tray containing round cavities an inch thick and 5½ inches in diameter, the curd is ladled into the cavities, and wooden discs are placed on top to apply pressure. The cheese is aged in caves dug into the hillsides, allowing yeast to grow on the surface. The cheeses are turned every other day, washed with whey and brine, and packaged with the wooden discs. Reblochon is soft to the point of runniness, and the rind is pink-white and covered lightly with yeast. The flavor is nutty, and the odor is not as intense as Limburger's or Brick's.

Smearing with Alcohol

Microbial growth on the rinds of some smear-ripened cheese is controlled by washing with locally produced alcohol instead of salt water. Époisses de Bourgogne (ay-PWOSS duh boor-GOYNE, named after its village of origin) is a strong candidate for the most pungent cheese on Earth (though it is not banned on public transportation in France, despite stories to the contrary). It is coagulated with starter culture only, taking at least 16 hours. It is then ladled into forms, turned three times to drain the whey, removed from the forms, and sent to a cooler room for salting. The cheeses are dried for up to a week and washed daily with *B. linens* in brine. The final washes contain Marc de Bourgogne, the local brandy. Other alcohol-washed cheeses include Belgium's Herve Affiné à la Bière ("Herve matured with beer"), Scotland's Bishop Kennedy (washed with malt whiskey), France's Saint Vernier (wine-washed), and England's Stinking Bishop (washed with pear cider).

Trappist-Style

Trappist-style cheeses with powerful odors such as Maroilles (mah-WAHL, dating to tenth-century France) and Livarot (LEE-vah-roe, thirteenth century) originated with Trappist monks. Maroilles is stored for 10 days while *B. linens* begins to grow on the surface. It is then washed and brushed for 2–4 months to remove mold and allow the *B. linens* to flourish, changing the color of the rind to orange-red. Unsurprisingly, the nickname for Maroilles is *vieux paunt* ("old stinker"). Livarot is always encircled with five narrow bands of red bulrush, originally used to keep the shape of this cylindrical cheese. The bands resemble the five stripes on a French colonel's uniform, giving this orange-brown cheese the nickname "the colonel." Both of these varieties are soft and riddled with small eyes.

Other Smear-Ripened Cheeses

Munster is named after a village in Alsace, France, and is identical to Géromé, named after the village of Gérardmer in Lorraine. This variety also originated with monks (Munster is from the Latin *monasterium*, monastery). The milk comes from cows native to the Vosges Mountains, and the curd is not washed or kneaded before being placed in forms. The cheese is brine-washed every other day for 2–3 months and has a soft, sticky, powerfully pungent rind and mild interior. Munster is not the same as the bland Münsters from Denmark and Germany.

Pont-l'Évêque (pawn le VECK, named after the village where it originated in the twelfth century) is said to be the third-most-popular cheese in France behind Brie and Camembert. The milk is coagulated at 95°. The curds are then cut into blocks, and whey is drained by placing the blocks in cloth bags and then by pressing in square metal boxes on straw mats. The cheeses are held in a room at 68°–77°. *Geotricum candidum* mold initially grows on the surface, but is controlled by daily washing with a salt solution. The cheeses are removed from the boxes within five days and brushed with coarse salt daily over the next 10 days. The salt draws out moisture and creates a rind, allowing to cheese to ripen properly. The cheese is aged 4–6 weeks and packed in square wooden containers measuring up to 8¼ inches on a side. The top rind bears the cross-hatching from the straw.

Serra da Estrela, a Portuguese sheep's milk cheese made with cardoon thistle rennet and named after the mountain range where it originated, is subjected to high humidity for two weeks to allow yeast to grow. The smear is then removed, and the cheese is transferred to a low-humidity room to dry the surface and allow the formation of a rind.

Teleme (TELL-uh-may) was developed by Greek immigrants to Northern California after World War I. It is formed in squares weighing up to 6 pounds, dusted with rice flour to absorb moisture, and ripened for two months. At that point, the surface is flecked with green and black mold, and the center is almost liquid.

Tilsit, or Tilsiter, invented in the 1800s by Swiss cheesemakers who settled near the East Prussian town of Tilsit (now Sovetsk, Russia), is popular in Germany and surrounding countries. The yeasts on the surface come from the environment, and young German Tilsit is smeared with growth from older Tilsit. Danish Tilsit is milder, but both versions have a brown-orange rind and a yellow interior sprinkled with small eyes. The pH on the surface increases to 7.5 during the first two weeks, allowing salt-tolerant bacteria to grow.

Aroma

The sense of smell, as discussed in the previous chapter, is an important component of flavor. Smear-ripened cheeses are the most odiferous on the planet because of the compounds generated on the surface, though the interior is usually mild. Not only is the casein degraded, but extensive breakdown of the lipids generates pungent short-chain fatty acids. Butyric, caproic, caprylic, and capric acids are responsible for the fatty, rancid, sweaty aromas in these varieties. These compounds are major contributors to foot odor and might attract mosquitoes (Box 8.2). Methionine in these

BOX 8.2
FOOT ODOR AND MALARIA

> *"Mr. Bloom ate his strips of sandwich, fresh clean bread, with relish of disgust pungent mustard, the feety savour of green cheese."*
> —JAMES JOYCE, *Ulysses*, 1922

The aroma of sweaty feet is generated by *Brevibacterium epidermidis*, which resides between the toes and is closely related to the *Brevibacterium linens* used for making smear-ripened cheese. *B. linens* metabolizes amino acids and lipids in cheese, *B. epidermis* does the same with dead skin, and both convert methionine into odiferous methanethiol. Limburger and foot perspiration also contain plenty of short-chain fatty acids. These facts carry implications for trapping mosquitoes. The *Anopheles* mosquito, dreaded for being the most effective carrier of the microorganism that causes malaria, comes in several species. Females of one of these species, *Anopheles gambiae* subspecies *Giles*, prefer to bite human feet and ankles, and are attracted to their odor. Washing with a bactericidal soap removes the attraction. Dutch researchers, who noted that their word for foot odor, *tenenkaas*, means "toes-cheese," theorized that Limburger could be used as an attractant for *Anopheles gambiae*. A homemade trap containing warm water (for providing heat and moisture) and Limburger was tested in Western Kenya and collected about two *Anopheles gambiae* per day, as opposed to one per day with cold water and Limburger, and none at all with Limburger and no moisture. It seems that Limburger cheese has the potential to serve as a medium for trapping dangerous mosquitoes.

cheeses breaks down into dimethyl disulfide and other sulfur-containing compounds (Box 7.1). Indole (Box 5.2), isovaleric acid (Box 5.4), and phenol (Box 6.1) also contribute to the aroma of Limburger.

There have been studies on the primary flavor compounds in some of the smear-ripened cheeses. In addition to the compounds above, Pont-l'Évêque contains several ketones, which will be discussed in the next chapter, and Livarot and Maroilles generate 2-phenylethanol (Box 6.1). Livarot also has *meta*-and *para*-cresol (Box 6.1), accounting for its pungent barnyard odor.

BOX 8.3

ESTERS

Cheese contains a number of esters. Box 3.1 showed the structure of ethyl octanoate, which has an aroma of apricot. Methyl thioacetate, mentioned in Box 7.1, is the key ester in Limburger.

Other important esters include the following.

Ethyl acetate has an aroma of pear and apple. It is the most common ester in wine.

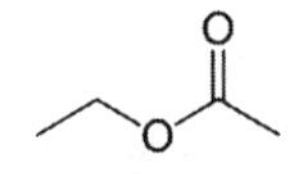

The main carbon chain of *ethyl butanoate*, shown here, has two more links than ethyl acetate. Ethyl butanoate imparts a pineapple odor and is often added to orange juice to enhance aroma. *Ethyl hexanoate* contains six carbon atoms in the main chain and has the odor of pineapple and banana. *Ethyl decanoate* has ten carbon atoms in the main chain and has an aroma reminiscent of fruit or wine.

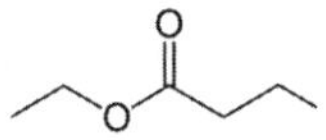

Ethyl-3-methylbutanoate has been described as having a fresh cheese aroma.

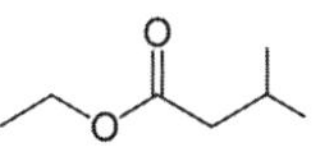

Isopentyl acetate (or isoamyl acetate) has a strong banana flavor and is sometimes called banana oil. It can also have a pear aroma.

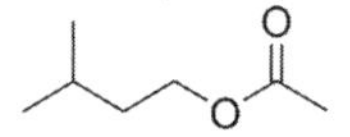

Phenyl ethyl acetate has a rose-like flavor and is found in fruit, whiskey, and wine.

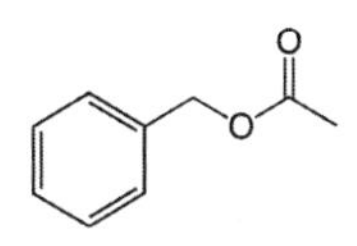

The aroma of cheese is subjective, and some people are repelled by it. Several people have told me that they stay away from Parmesan because "it smells like vomit." Butyric acid is a noted odorant of Parmesan and is generated in your stomach when the hydrochloric acid there reacts with food. Some compounds created in cheese are also found in decaying organic matter, which is not surprising since both processes involve decomposition of fat and protein. Humans are naturally repulsed by spoiled food, and sensitive individuals consider aged cheese to be in this category.

Esters

Esters arise in cheese when alcohols from amino acid or lactose metabolism react with fatty acids. A wide variety of enzymes are responsible for the creation of esters in cheese, and their mechanism of formation is largely unknown. Esters containing twelve or fewer carbon atoms are fragrant compounds, typically with fruity aromas, and are commonly used as artificial flavors and fragrances. The smaller esters are perceived in cheese at the parts-per-million range, which is ten times lower than the corresponding alcohols. They may also lessen sharpness and bitterness. Many of the important esters found in cheese are described in Box 8.3. Other esters not commonly observed in cheese carry the characteristic aroma of the fruit in which they are found: methyl butyrate is in apples, isobutyl formate is in raspberries, and octyl acetate is in oranges.

Esters are also formed in osmophore cells in flowers, where they arise from reactions of acids and alcohols. These sweet-smelling compounds, such as benzyl acetate in jasmine, draw pollinating insects to the flower. Humans are conditioned to like these odors, though there might be a genetic component to our attraction to them.

Esters and ketones are important contributors to the aroma of interior mold cheeses, and we turn to those next.

Roquefort should be eaten on one's knees.

—ALEXANDRE BALTHAZAR LAURENT GRIMOD DE LA REYNIÈRE, French gourmet writer (1758–1837)

9

INTERIOR MOLD CHEESES, KETONES, AND MICROBIAL STRAINS

ROQUEFORT IS ONE of the most intensely flavored cheeses and has moved people to rhapsodize about it. Pliny the Elder wrote favorably about a similar variety in his *Natural History* in AD 79, and it probably dates back many years before that. Interior-mold cheeses are skewered with stainless steel or copper needles, introducing oxygen and allowing the mold to thrive inside the channels created, and in the fissures between the curd pieces. The mold dominates proteolysis and lipolysis during ripening, metabolizing amino acids and fatty acids and creating spores, forming the blue-green veins. The pH of the cheese increases at the same time, affecting texture and flavor. Yeasts can also find their way into the cheese and contribute to ripening.

The "big three" interior mold cheeses are Roquefort, Gorgonzola, and Stilton. *Lactobacillus lactis* starter culture is used, sometimes with Leuconostoc species that create CO_2 gas that helps open the structure and give the mold room to grow. Indigenous *Penicillium roqueforti* molds create the veins, though this genus of fungi is better known for another notable activity: saving lives (Box 9.1).

Roquefort

According to legend, a shepherd boy left his meal of bread and sheep's cheese in a cave to chase after a pretty girl; when he remembered the meal weeks later, the cheese displayed veins of mold. He ate it anyway, liked it, and Roquefort cheese was born.

BOX 9.1
PENICILLIN

Bread with blue mold on it was used to treat open wounds in Europe in the Middle Ages. Nobody knew at the time that the mold contained species of Penicillium (from the Latin *penicillus*, "paintbrush," referring to its appearance under the microscope), which secrete a compound that interferes with production of bacterial cell walls. The bacteria die as a result, making Penicillium extract an effective treatment for many infections. But the mechanism of disease transmission was unknown until the late nineteenth century, and people died of infectious diseases in numbers that would be alarming to us today.

The use of penicillin as an antibiotic for treating bacterial infections was pioneered by Alexander Fleming, a bacteriologist working at St. Mary's Hospital in London. Sixteen years after the discovery, he told a story that is now famous: in the fall of 1928 he noticed that a staphylococcus culture in a petri dish in his laboratory was killed by *Penicillium notatum* spores that had floated in through an open window while he was on vacation. Fleming realized that the mold could be useful instead of a nuisance, identified the antibacterial agent, and named it penicillin. In reality, the events could not have taken place as he remembered, since no bacteriologist would have an open window through which contaminants could drift in (his small laboratory window was never opened anyway), and the Penicillium would have overwhelmed the culture in the dish and not just established a few colonies, as he later recalled. The Penicillium probably came from the lab below Fleming's (where the same strain was grown), and he performed an experiment on it as part of a study on something else (he mentioned this experiment in his notebook weeks after his vacation, but wrote nothing about the staphylococcus in the petri dish). At any rate, Fleming set out to isolate penicillin in a form and quantity suitable for administering to people, but failed. Fortunately, Howard Florey and Ernest Chain at England's Oxford University succeeded in converting penicillin into a stable powder in 1940.

The next challenge was scaling up the process. Great Britain was involved in World War II by then and did not have the resources to pursue penicillin manufacture, so the scene shifted to America. Originally, the mold was grown as a mat atop a nutrient solution. Difficulties with contamination and isolation resulted in enough penicillin for treatment of just one case by March 1942. Scientists at the USDA's Northern Regional Research Laboratory (now the National Center for Agricultural Utilization Research) in Peoria, Illinois, began pumping air into deep vats containing *P. notatum* suspended in corn-steep liquor, a non-alcoholic byproduct of wet-milling corn. This process was far more effective, and in July 1943, the War Production Board asked for mass-production of penicillin to begin as soon as possible. The Peoria lab began a search for the Penicillium species that would grow best in the huge vats. Since Penicillium grows in soil, U.S. Army pilots around the world sent in dirt samples to be tested. Peoria residents and lab employees were encouraged to donate moldy household objects. One lab technician, Mary Hunt, was assigned the task of visiting area groceries to locate spoiled

BOX 9.1 *(continued)*

food, earning her the nickname "Moldy Mary." The species that grew best, *Penicillium chrysogenum*, was found by Hunt on a cantaloupe from a Peoria market, and was used in the research. The process was successfully scaled up by several pharmaceutical firms, and the United States had 2.3 million doses in time for the Normandy invasion on D-Day, June 1944. The drug was available to the general public by war's end. Fleming, Florey, and Chain never patented penicillin and earned no royalties on it, though they did share in the 1945 Nobel Prize in Physiology or Medicine.

Penicillin was credited with saving countless lives and limbs during the war as it prevents gangrene, a common battlefield infection, and destroys other bacteria that cause disease. It is still widely used as the first line of defense against many dangerous bacteria.

Roquefort is made by seven companies in and around Roquefort-sur-Soulzon, France. If a similar cheese comes from outside that area, it must be called something else, such as *blue cheese*. The town is on the side of Le Cambalou hill, which contains natural limestone caves. The cheese comes from unpasteurized, unhomogenized sheep's milk and is only made from December to July, when the sheep lactate (the rocky terrain allows for sheep grazing and little other agriculture). Roquefort is coagulated with locally obtained lamb's rennet at 85° and held without stirring for two hours. The curd is cut, stirred, drained, and ladled into perforated cylindrical metal forms where *Penicillium roqueforti* is added. Roquefort and other interior mold cheeses are not pressed, so small holes and cracks remain in the curd and provide space and oxygen for the mold to develop. The cheeses are turned frequently for a week in a room at 50° and 98 percent relative humidity before being sent to the caves, where the temperature is 43°–46° and humidity is also about 98 percent. The air currents that pass through the rock fissures help maintain the cave environment. The cheeses are covered with NaCl and skewered, and the surfaces are scraped after a week. They are then placed on oak shelves and ripened there for three months before being transferred elsewhere in the cave for at least three more months. The pH of the cheese increases to 6.4 as the mold converts amino acids to ammonia and other products.

P. roqueforti for use in Roquefort cheese is traditionally obtained from homemade rye bread that is over-baked and allowed to deteriorate for a month in the caves. The coat of mold that develops on the bread is powdered and sprinkled on the cheese, though nowadays the mold is also grown in laboratories from strains recovered from the caves. Some 600 strains of *P. roqueforti* are known to exist, and the damp and drafty caves are perfect environments for their growth. After a month of storage in the caves, the cheeses are wrapped with aluminum foil to prevent rind formation and further mold growth on the surface. The cheese is crumbly with green-blue veins; sheep's milk contains no carotenoids, so the cheese apart from the veins is white. A wheel weighs about six pounds.

Gorgonzola

Gorgonzola is named after an Italian town near Naples and was recorded as being made there in AD 879. The legend states that a cheesemaker left work early to visit his girlfriend, leaving a batch of curd hanging overnight. The next day, he added it to the new milk and made cheese out of it to cover his mistake. He discovered a green mold in the center of the cheese a few weeks later, sampled it, repeated the procedure, and a new variety was created. The milk originally was supplied by cattle that were rested in and around Gorgonzola while being driven between Alpine and lowland pastures. The cattle produced much more milk than the town needed, and cheese was made out of it. The cattle were weary from their journey, which is why a similar cheese is called Stracchino (from *stracca*, "tired").

Gorgonzola is made by roughly 40 companies. The process begins when starter culture, rennet, and *P. roqueforti* are added to whole milk at 86°. The curd is cut into pieces measuring ¾–1 inch, stirred, and drained at 64° for 10 hours. The traditional *piccante* ("spicy") or *naturale* Gorgonzola is made from two milkings, with cool curds from the previous evening's milk being loosely sandwiched inside warm fresh curds from that morning's milk. The evening curds are held in cloth bags to drain the whey, removed, and placed in the center of the forms lined with the fresh curd. The *dolce* ("sweet") version comes from one milking and is much more common. The curd is salted and held for 3–4 days in "purgatory," a room at 68°–72° and 95 percent relative humidity. The cheese, shaped into 26-pound wheels with a diameter of 12–14 inches and a height of 9 inches, is pierced after 12 and 20 days and stored 3–6 months. Before shipping, the wheels are bisected horizontally for easier handling. The pH is 5.15–5.30.

When a building housing the London Stock Exchange was completed in 1884, it was nicknamed Gorgonzola Hall because of the greenish veins contained in the interior marble walls.

Stilton

British author Gilbert Keith Chesterton, who once said "the poets have been mysteriously silent on the subject of cheese," tried to rectify the situation by writing *Sonnet to a Stilton Cheese* in 1912. Though not as world-renowned as Roquefort, Stilton is

England's most famous interior mold cheese. It is no longer made in Stilton, Cambridgeshire, but by five companies in the counties of Derbyshire, Leicestershire, and Nottinghamshire, which are to the north and west. The cheese was apparently developed in the seventeenth or early eighteenth century and became popular when a Stilton inn began to sell it to travelers. We have no evidence of a lovesick boy's inventing the cheese after chasing a girl.

Seventeen gallons of locally produced pasteurized milk are needed to make a 16-pound wheel of Stilton. The milk is cooled to 86°, and starter and *P. roqueforti* are added. The curd forms after rennet is added and settles at the bottom of the vat as the whey is drained overnight. The curd is cut, milled, dry salted, and placed in cylindrical hoops, and the whey drains for 5–6 days without pressing. The temperature is held at 79°–86° and 90 percent relative humidity to allow the starter bacteria to generate acid. The hoops are removed, and the cheese is stored at 55°–59° and 90 percent relative humidity for a few weeks. The cheese is skewered after four weeks and is ready for sale after nine weeks. The pH of the cheese is a relatively low 4.7 at the start of ripening, resulting in a crumbly texture. A week or two later, the pH begins to climb as a result of proteolysis by the mold, eventually reaching 5.5. Suffolk Blue and Yorkshire are similar cheeses manufactured in those counties. Stilton may also be made without the mold, and is then called White Stilton.

Danablu

Production of Danablu, or Danish blue cheese, began in 1927. Danablu is made from cow's milk that is not pasteurized but instead *thermized* by heating it to 140°–150° for 15 seconds. This milder treatment allows the indigenous lipases in milk to remain active. Unusually for cheese, the milk is also homogenized, which promotes activity by lipase enzymes and enhances the flavor. Chlorophyll or titanium dioxide is sometimes added to whiten the product and make it appear like sheep's milk cheese. Starter and rennet are added to milk at 91°, and the coagulation takes place over one hour. The curd is cut, and 90 minutes later the whey is drained and the curd is broken up. One pound of NaCl is sprinkled on for every 100 pounds of curd, followed by *P. roqueforti* powder. The curd is placed in hoops and held overnight at 70° as the pH decreases to 4.5. NaCl is then rubbed all over the cheese, preventing the starter bacteria from surviving. The mold is salt-tolerant, however, and will begin to appear within 10 days. The CO_2 that had been generated from the starter creates small openings in the curd that enable additional growth of mold. The cheese is pierced the next day and then stored at 55° and 95 percent relative humidity for 2–3 months. Like Stilton, the pH starts at 4.6–4.7 and increases with storage.

Maytag Blue

Maytag Blue has been made by hand in Newton, Iowa since 1941, after the procedure was developed as an American version of Roquefort. The company is owned by the descendants of the founder of the Maytag appliance company. The milk is heated to 90°–92°, homogenized, and coagulated with rennet at that temperature. The curd is cut after 30–45 minutes and ladled onto cheesecloth that is laid across the top of the vat. *P. roqueforti* and salt are sprinkled on the curd, which is rolled back and forth on the cheesecloth to drain some of the whey, resulting in a thick, crumbly texture in the final product. The cheese is skewered after three days to let in oxygen and promote mold growth. The cheese is aged for six weeks in caves dug into the side of a hill at the nearby dairy farm, after which outside mold is removed and the cheese is coated with wax and aged a few more weeks. When the cheese is ready, it is pierced again, the wax is removed, and the product is packaged in aluminum foil.

Ketones

Ketones are common industrial chemicals that are often used as solvents in the explosives, fragrance, paint, and textile industries. The simplest and most important ketone in the marketplace is acetone, nail polish remover. In nature, ketones are found in some sugars such as fructose, and are produced in the body through breakdown of amino acids and fatty acids. When the body runs out of glucose in the liver for energy, it begins to break down stored fat into ketones. Cortisone is a well-known synthetic ketone.

Many compounds contribute to the odor of a particular cheese variety, and much of the aroma of blue cheese is traced to ketones, specifically methyl ketones such as 2-undecanone (Box 9.2). Ketones arise in cheese when triglycerides undergo lipolysis. One half to three-quarters of the total aroma compounds in this class of cheese are due to methyl ketones. In Roquefort and Gorgonzola, 2-heptanone is the most abundant ketone, followed by 2-nonanone. The key ketones in Stilton are 2-heptanone, 2-butanone, and 2-pentanone. Enzymes break down ketones into alcohols such as 1-octen-3-ol (mushroom flavor) and 2-heptanol (herbaceous flavor).

Interior mold cheeses generate large amounts of free fatty acids: Danablu and Roquefort each have 3.2 percent, whereas Parmesan contains 0.5 percent and Cheddar has 0.1 percent. These fatty acids would ordinarily produce a rancid flavor, but they are neutralized as the pH increases in interior mold cheeses and some are converted to methyl ketones.

BOX 9.2
KETONES

As we saw in Chapter 5, nearly all fatty acids in cheese have an even number of carbon atoms. When a microorganism attacks a fatty acid, its lipases often remove oxygen and one carbon atom to make carbon dioxide, CO_2. Therefore, most ketones in cheese have odd-numbered carbon chains. The oxygen atom in the ketone is frequently attached to the second carbon atom in the chain, making it a *methyl ketone*. A common compound in cheese, 2-undecanone, has 11 carbon atoms (hence the *undec-*, the prefix for 11) and looks like this:

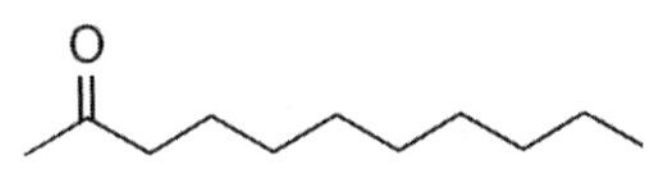

The C=O group is at the second position, accounting for the "2" prefix. Similarly, 2-heptanone has seven carbons, and 2-nonanone has nine. These two ketones are primarily responsible for blue-cheese flavor. 2-Undecanone and 2-pentanone (with 5 carbons) are described as floral or fruity. Ethyl ketones with an even number of carbon atoms and commonly found in cheese are 2-butanone (also known as methyl ethyl ketone, with a solvent odor) and 2-hexanone (blue-cheese aroma).

We mentioned 1-octen-3-ol in Box 6.1. A similar compound is 1-octen-3-one, a ketone that also has a mushroom aroma. It has a C=O group at the third position:

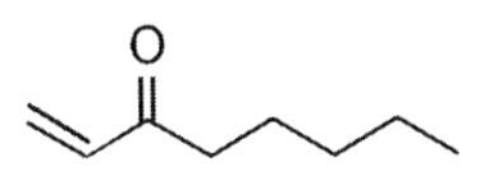

Some ketones have more than one C=O attached. 2,3-Butanedione has two carbonyl groups:

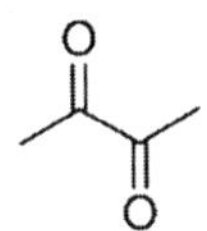

2,3-Butanedione, also known as diacetyl, has a strong buttery aroma. If you buy a sealed package of a cheese that hasn't been aged long, such as cottage cheese or mild Cheddar, put your nose close to the container and open it. The odor that first hits you is diacetyl. Another odorant is 3-hydroxy-2-butanone, also known as acetoin. It is identical to diacetyl except it has a hydroxy group instead of one of the carbonyls. Acetoin has a buttery or sour-milk odor. Related compounds (though not ketones) are 2,3-butanediol, which has hydroxy groups in place of both carbonyl groups, and pyruvate, shown here:

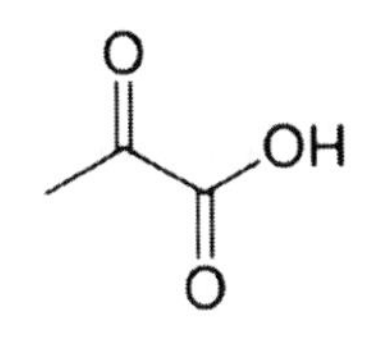

BOX 9.3
CLASSIFYING DAIRY ANIMALS

Taxonomy is the science of identifying living things and fitting them into a biological classification. From kingdom down to family, the most common dairy animals are classified as follows:

Kingdom:	Animalia (animals)
Phylum:	Chordata (have spinal cords)
Subphylum:	Vertebrata (have backbones)
Class:	Mammalia (mammals)
Order:	Artiodactylia (have hoofs with even numbers of toes)
Family:	Bovidae (cloven-hoofed ruminants)

The split comes with "subfamily," which is inserted to distinguish size:

Subfamily:	Bovinae (bovines, large-sized bovids)
	Caprinae (caprines, medium-size bovids)

Among the bovinae:

Genus:	*Bos*	*Bos*	*Bos*	*Bubalus*
Species:	*taurus*	*grunniens*	*primigenius*	*bubalus*
Subspecies:			*indicus*	
Common name:	Cow	Yak	Zebu	Water Buffalo

Among the caprinae:

Genus:	*Capra*	*Ovis*
Species:	*aegagrus*	*aries*
Subspecies:	*hircus*	
Common name:	Goat	Sheep

We mentioned reindeer and moose in Chapter 1. They belong to the Cervidine family, which consists of deer and related animals.

Genus:	*Rangifer*	*Alces*
Species:	*tarandus*	*alces*
Common name:	Reindeer	Moose

Adjectives describing animals are taken from the family or generic names. Cows, yaks, and zebus are bovines; water buffalo are bubalines; goats are caprines; sheep are ovines; reindeer are rangiferines; and moose are cervines.

BOX 9.4
ANOTHER USE FOR FUNGUS

Microorganisms can be used for various purposes, not just fermenting food. For instance, a team in Germany fashioned a cleaning agent using *P. roqueforti* mounted on plastic. The device consisted of a three-layer sandwich: a plastic base layer used as a support, the living layer consisting of the fungus habitat, and a porous cover membrane for protecting the living layer from its surroundings while allowing gases and nutrients to pass through. The cover also protects the human user from the fungi. Under proper environmental conditions, the fungus remains in a nearly dormant state for weeks at a time until coming in contact with a nutrient source such as bacterial contamination or spilled food. The *P. roqueforti* consumes the nutrient while staying in its habitat and then reverts to its near-dormant condition. This arrangement can be used in a hospital setting since the cover allows the living layer to be active even if in contact with disinfectants or detergents. The research team speculates that antibiotic, self-sterilizing units will eventually be developed from this technology.

Species, Subspecies, and Strains

Every organism is classified biologically in seven major ranks: kingdom, phylum, class, order, family, genus, and species (Box 9.3). The last two are the generic and specific names. The generic name is always capitalized; the specific name never is; and both are italicized. The kingdoms include animals, plants, and, importantly for cheesemaking, fungi and bacteria. The fungus *P. roqueforti* is classified as phylum Ascomycota (distinguishing characteristic: forms spores in sacs), class Eurotiomycetes (distributes spores in a certain way), order Eurotiales (blue and green molds), family Trichocomaceae (processes dead organic matter), genus *Penicillium* (brush-like appearance), species *roqueforti* (grows in 0.5 percent acetic acid). Classifications change when new information comes to light. Scientists used to think that *P. glaucum, P. gorgonzola, P. stilton*, and several others were different from *P. roqueforti*, but they are now classified as the same species. Scientists are also finding new ways to utilize *P. roqueforti* (Box 9.4).

A common bacteria used in starter cultures, *Lactococcus lactis*, has this classification: phylum Firmicutes (strong cell walls), class Bacilli (certain compounds in cell walls), order Lactobacillales (metabolizes lactic acid), family Streptococcaceae (grouped in chains), genus *Lactococcus* (spherical or oval cells), species *lactis* (found in milk). Two subspecies are used in cheesemaking, *lactis* and *cremoris*, which differ in the conditions in which they thrive: *cremoris* cannot grow in 4 percent NaCl or at 104°, whereas *lactis* can do both. Cheesemakers use mixtures of the two subspecies depending on the rate of acid formation and level of proteolysis they wish to attain.

A "strain" is a group of organisms capable of interbreeding, and is defined as a population of genetically identical cells that is a variant of a bacterium, fungus, or virus. Strains are part of the same species but have some differences (though not enough to be a separate variety). In one study, milk, curd, and cheese were collected in seven major cheesemaking regions of France, and strains of *Geotrichum candidum* fungus that colonizes most surface mold cheeses were isolated. Of the 64 strains that were obtained, 55 could metabolize lactic acid; nine could not. Some identical strains were found in places that were far apart geographically, and some facilities had several strains present. It appears that certain strains develop and thrive depending on temperature, NaCl content, and so forth. The diversity of microorganisms contributes to the diversity of flavor in cheese. They also contribute to texture development, which we will see in the next chapter.

Poets with whom I learned my trade,
Companions of the Cheshire Cheese
—WILLIAM BUTLER YEATS, *Responsibilities and Other Poems*, 1916

10

CHEDDARED CHEESES, ALDEHYDES, AND TEXTURE

YEATS WAS REFERRING to Ye Olde Cheshire Cheese, a London pub that has been operating continuously under that name since the 1600s. Cheddared cheeses such as Cheshire and Cheddar were created in Great Britain centuries ago and come under the category of British Territorial cheeses, which are named after the place where they originated. These cheeses are hard, semi-hard, or crumbly, and are distinguished from other types by the cheddaring step, where curd slabs are stacked atop each other. During World War II, most of the cheesemilk in the United Kingdom was diverted to making Cheddar, and production of most other varieties was curtailed. "Government Cheddar" was rationed along with other food due to wartime shortages. The restriction on cheeses other than Cheddar was not lifted in the United Kingdom until 1954, and some varieties and many cheesemakers had disappeared by then, never to return.

Cheddar

Cheddar cheese is the world's most popular variety, and is certainly the most studied—more than 1,400 scientific publications mention it in their titles. It was first produced in Cheddar, Somerset County, in the twelfth century. It is no longer made there, but some is still aged in caves at Cheddar Gorge, a picturesque limestone gorge near the village (Box 10.1). Some 333 farms in Southwest England made Cheddar as late as 1939, but the effects of the war and the spread of industrial dairies reduced the number of farms to 32 by 1974. Manufacture begins when the milk is heated to 88° and starter culture containing *Lactococcus lactis* and Leuconostoc is added (*Lactococcus lactis* is so prominent in Cheddar, Colby, and Monterey Jack cheese manufacturing that the Wisconsin state legislature voted it the official state microbe in 2010). Calcium chloride and rennet are added after 15–30 minutes, and the milk is then stirred for 5 minutes. Coagulation occurs within 25 minutes, and the curd is cut with quarter-inch knives into pieces the size of a pea. After another 5–15 minutes of stirring, the vat is heated to 100°–102° with stirring for 30–45 minutes, and stirring continues for another 45–60 minutes after the temperature is reached. The process of cutting curd into small pieces and heating it over 100° is known as *scalding*, and allows for uniform expulsion of whey. When the curd pH reaches 6.1, the whey is drained and the cheddaring process begins. Slabs of curd around 6 inches wide are sliced along the sides of the vat and turned over every 10 minutes. The slabs are stacked two high and then four high during a 90-minute period, forcing out whey. As the pH drops to 5.2, the pressure, heat, and action of the starter bacteria cause the curd to mat into a solid mass with a texture similar to that of cooked chicken breast. The curd is sliced and sent through a milling machine to chop it into small pieces, which are then dry-salted in several applications. The curd is then let alone or "mellowed" while the salt dissolves in the water on the surface of the pieces. The curds are transferred into metal forms lined with cheesecloth and pressed. The aging ranges from three months to over two years in a 42°–50° room with a relative humidity under 80 percent. Nonstarter lactic acid bacteria, mostly Lactobacillus species, dominate during most of the ripening period and impart desirable flavors. Some strains are added as adjunct cultures to improve flavor quality, though some of these flavors by themselves are not exactly pleasing (Box 10.2).

Traditional farmhouse Cheddar wheels are entirely handmade, cylindrical, and wrapped in muslin. In large-scale production, most steps are automated, and the cheese is in the form of plastic-wrapped blocks. Cheese vats are started every half-hour, and draining, salting, and mellowing take place on a conveyor belt. Cheddaring is performed by continuously passing curd into a large tower where the weight of the curds above squeezes out whey. Another tower is used to press the curd into blocks, which are guillotined into the desired size and vacuum-packaged. An acceptable Cheddar cheese can be made with a pH level anywhere from 5.0 to 5.4, making it a relatively easy cheese to manufacture compared to varieties with tight limits. Cheddar,

BOX 10.1
CHEDDAR GORGE

Cheddar Gorge is a popular tourist attraction and considered one of England's greatest natural wonders. The gorge was formed by a surface river cutting into the limestone over the past 1.2 million years. Many caves are there, and the two largest, Cox's Cave and Gough's Cave, are open to the public. The gorge is home to several rare or unique species of plant life.

Great Britain's oldest complete human skeleton, dating back to around 7150 BC, was found in Gough's Cave in 1903. Nicknamed "Cheddar Man," he was an adult male who was apparently killed—the side of his skull has a hole in it. In 1997, Oxford University researchers led by Bryan Sykes extracted mitochondrial DNA (mtDNA) from one of the molars. DNA carries all genetic information for an individual; mtDNA comes from cellular power plants, the mitochondria, and is inherited only from the mother. Cheddar Man's mtDNA was compared to scrapings from the cheeks of 20 local residents, and one local teacher, who was born 15 miles away in Bristol, was a match. He and Cheddar Man therefore both had a common female ancestor.

BOX 10.2
MEOW

"It's a Cheshire-cat," said the Duchess, "and that's why."
—LEWIS CARROLL, *Alice's Adventures in Wonderland*, 1865

2-Mercapto-2-methylpentan-4-one is both a sulfur-containing compound and a ketone, with this structure:

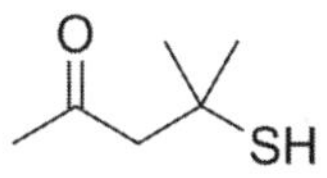

It has been found in cheese, black currants, and other foods. It is also a breakdown product of a longer molecule, felinine, an amino acid found in the urine of male domestic cats. Felinine is presumed to be involved in the production of pheromones, which the cat uses to mark territories in order to attract females and repel other males. Tomcats excrete four times as much felinine as female cats or castrated male cats. 2-Mercapto-2-methylpentan-4-one helps impart the characteristic odor to tomcat urine and is responsible for the catty odor that aged cheese sometimes has. In cheese it probably arises from the breakdown of methionine in casein and may be considered a desirable flavor in Cheddar. In fact, at least one noted manufacturer deliberately adds an adjunct culture that produces 2-mercapto-2-methylpentan-4-one in one of its Cheddar brands. Cheese tasters and graders will often say "meow" when they encounter this aroma in a product.

Stilton, Tomme de Savoie (Chapter 12), Mimolette (Box 16.5) and some other varieties are natural rind cheeses, meaning that their rinds are created during aging as the outer surface dries out (though industrial Cheddar no longer has rinds). These rinds are not washed, and no bacteria or molds are added to them (as opposed to the cheeses in Chapter 8), though microorganisms from the environment may grow on them. Cheese rinds are safe to eat as long as they are clean and have no cloth or wax attached.

Cheshire

Cheshire cheese was first mentioned by name in the late sixteenth century, though it may date to pre-Roman times, and is sometimes called Chester after the port from which it used to be shipped. It is now made in the English counties of Cheshire, Shropshire, and Staffordshire, and in the neighboring Welsh counties of Denbighshire and Flintshire. The Cheshire basin is an area of Great Britain featuring salt beds and a high amount of salt in its soil. The salt finds its way into pasture plants, imparting a unique flavor likened to sea spray or salt marsh. Milk is heated to 90°–95° and acidified quickly with large amounts of starter. The acidic conditions increase the rennet activity and the loss of colloidal CCP to the whey. After cheddaring, the curd is milled to a granular texture, and the pH at pressing is only 4.8, leading to a crumbly cheese. Cheshire is stored at 43°–46° for 6–9 months and the texture and flavor have been described as being midway between that of Cheddar and Feta.

In 1988, U.S. Customs asked our research group to help them with a problem: they had seized a shipment of English cheese described as "Cheddar," but suspected that it was actually Cheshire that was mislabeled to circumvent import quotas. The United States has restrictions on Cheddar imported from England, since that variety is made domestically, but none on Cheshire, which is not manufactured in America. We found that the best way to distinguish the two was through the use of rheology, the study of flow and deformation of matter. Our instrumental measurements showed that the body of the suspect cheese did not hold together nearly as well as that of genuine Cheddar of the same age, an indication that the cheese was really Cheshire. The importation rules were tightened up after that.

Red Leicester and Double Gloucester

Red Leicester (LESS-ter) was first made in Leicestershire in the seventeenth century and is similar to Cheshire. It is traditionally made in wheels weighing 22–44 pounds and is ripened at 50°–59° for 4–8 months. Annatto is added to provide its characteristic tangerine color, which originally came from beet or carrot juice. During World War II, Leicester was produced without the annatto, giving it a yellow color like most other cheeses'. When the dye was permitted again, manufacturers added the "red" in the name to distinguish it from the wartime variety. Annatto is used by some manufacturers to dye Cheshire as well.

Double Gloucester (GLOSS-ter), dating to the sixteenth century, is made from whole milk, and not partly skimmed milk as in Single Gloucester. Both are made with milk from the rare Gloucester cow. The curds are cooked at 95°–100° and are milled twice. Annatto is often added to Double Gloucester, and the cheese is aged 4–6 months, developing a rind. Double Gloucester is usually shaped in wheels weighing 8–18 pounds, and Single Gloucester comes in wheels weighing 7–12 pounds; this difference may also account for the adjectives "double" and "single." Fourteen-pound Double Gloucester cheeses are used every May in the Cooper's Hill Cheese-Rolling and Wake, where a wheel is rolled down a 50 percent gradient slope in Brockworth, Gloucestershire, and people try to chase after it. There are men's and women's races; in between, paramedics remove injured competitors and spectators. A cheese is awarded to the first person in each race to reach the bottom of the hill without dying.

Other British Varieties

Caerphilly (care-PHIL-lee) originated in the area around the town of Caerphilly, Wales, where a cheese sculpture now stands and an annual three-day Big Cheese festival is held. Caerphilly dates back to the 1830s and is the only traditional Welsh cheese. Mesophilic starter cultures and calf rennet are used to produce the curd, which is held at 90°–93° for an hour. The curd is cut into half-inch cubes using knives that reach the bottom of the vat, and the curds and whey are hand-stirred. The correct acidity is determined by pressing a hand on the curd: if the imprint remains, the whey is drained. The curd is cut again, piled into forms, and pressed for less than 30 minutes. The cheeses are held in brine for a day to enable rind formation and are ready for sale in just two weeks, though they may be aged for 3–4 months. The less extensive pressing results in a higher moisture content and a softer, buttery texture. Caerphilly forms a gray rind, which may pick up mold growth. Welsh miners were said to have preferred this variety since the rind kept the dirt from their hands off of the cheese and the extra salt helped replenish what they lost through perspiration.

Dunlop is the indigenous hard cheese from Scotland, dating to the late 1600s. Production stopped during the war and has been revived by artisan cheesemakers using Ayrshire cows. It is similar to Cheddar, being aged 6–12 months to prevent buildup of acidic flavor. Derby (pronounced "DAR-bee") is also similar to Cheddar, but the curds are softer and the cheese has a loose texture. Authentic Sage Derby is made by adding chopped sage to the curds, giving the cheese a green marbling.

Wensleydale, a blue-veined cheese made for eight centuries until World War II, was revived in the early 1950s as a variety similar to Cheddar and Caerphilly. A small amount of the blue version is still made, but the rest is white with an aging time of a few weeks.

Locally obtained milk for Yarg Cornish cheese is coagulated, drained, cut into blocks, milled, brined for 24 hours, and allowed to dry for 48 hours. Edible nettle leaves are then applied by hand over the entire cheese surface, causing breakdown of the rind; the interior is crumbly like Caerphilly. The rind turns gray-green over the three-week ripening period as the nettles impart a mild mushroom flavor. Yarg is "gray" spelled backward, after the surname of the couple who developed this variety in the 1970s.

French Varieties

Cantal and Salers are similar cheeses from the Cantal Mountains in France and are made from milk of Salers cows. Cantal is mass-produced from milk of hay-fed cows from November to April, and Salers is artisanally made from pasture-fed cow's milk during the other months. Salers is the only major French cheese that is entirely farmhouse-made. Both are cheddared with the curd being cut into half-inch cubes, pressed, and rested for eight hours. The curd is then ground through a milling machine, salted, packed into forms, and pressed a second time. They come in the form of 80-pound wheels with thick rinds.

Aldehydes

Aldehydes tend to be transient compounds in cheese, breaking down quickly into alcohols or acids. Straight-chain aldehydes (those without branched chains) are usually formed by enzymatic degradation of fatty acids, and can produce highly undesirable flavors when their concentration is too high. The branched-chain aldehydes originate from amino acid breakdown, and do not have to involve enzymes. A common aldehyde in Cheddar, methional (Box 7.1), imparts a boiled potato aroma and 3-methyl butanal (Box 10.3) is also important. Three branched-chain aldehydes combine to create nutty flavors in Cheddar.

BOX 10.3
MORE ON TEXTURE

Rheology (from the Greek word for flow) is the study of flow and deformation of matter and was pioneered by Isaac Newton. Tribology (from the Greek for "I rub") is the science of interactive surfaces in motion and was first studied by Leonardo da Vinci. Both are involved in food texture because the eating process changes from deformation to lubrication before swallowing. Teeth, tongue, palate, and saliva all attack the food, altering it from a chunk up to an inch in size to pieces less than a millimeter across. In the process, the food is bitten with the front teeth, transported toward the molars, accumulated into a mass, and swallowed. The primary and secondary mechanical properties listed above cover the initial chewing of a food and are rheological. The table below shows the sensations during the eating of cheese that are categorized under rheology and tribology.

Parameter	*Descriptors*	*Definition*
Particle size, shape, orientation	Gritty, grainy, fibrous	Feel of food particles in mouth
Juiciness	Juicy, watery	Amount of water relative to solids
Fattiness	Greasy, oily	Amount of fat relative to solids
Creaminess	Creamy	Perceived smoothness of dairy foods
Smoothness	Smooth, rough	Force required for moving surfaces relative to each other

Fattiness is an important factor when consuming cheese, which has a substantial fat content, since fat and oil have been found to remain on the tongue as a film or as droplets a half a minute after the food is swallowed. One sensation that is categorized solely under tribology is astringency, the feeling inside the mouth of drying out, which we mentioned as a flavor factor in Chapter 7. Astringent compounds such as certain amino acids in cheese and tannins in wine combine with salivary proteins, resulting in an increase in friction inside the mouth.

TABLE 10.1

Mechanical properties of cheese

Parameter	Descriptors	Definition
	PRIMARY MECHANICAL PROPERTIES	
Hardness	Soft, firm, hard	Force required to compress cheese between molars
Cohesiveness	(See secondary properties)	Structural strength of the cheese
Springiness	Plastic, elastic	Degree to which cheese returns to original shape once it has been compressed between the teeth
Adhesiveness	Sticky, tacky, gooey	Force required to remove the material that adheres to the palate during eating
	SECONDARY MECHANICAL PROPERTIES	
Fracturability/ brittleness	Crumbly, crunchy, brittle	Force with which cheese crumbles, cracks, or shatters
Chewiness	Tender, chewy, tough	Work required to masticate the cheese until it can be swallowed

Texture

The texture of a cheese may be considered as important as its flavor. Consumers will not accept cheese, or any food, if it doesn't produce the expected sensation when they break off a piece and chew it. Even if the flavor is correct, people will shun a food that "feels wrong" to them. Texture is even more important in cheese when it used as an ingredient in another food. The cheese on a cheeseburger contributes to the overall flavor of the sandwich but becomes objectionable if it is rubbery or tough.

BOX 10.4
ALDEHYDES

The simplest aldehyde in cheese is acetaldehyde, with the structure at right:

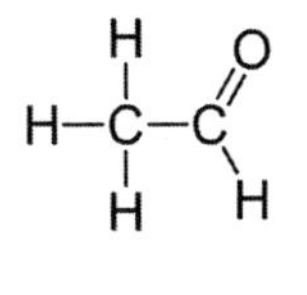

or just

when the C's and H's are removed.

Other simple straight-chain aldehydes are hexanal (6 carbon atoms), heptanal (7 carbons), octanal (8 carbons), nonanal (9 carbons), and decanal (10 carbons). All have aromas described as green, fatty, herbaceous, and soapy. The most common aldehyde in cheese is nonanal, with this structure:

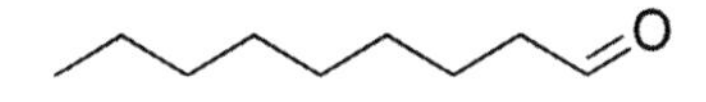

The major aromatic aldehyde in cheese is phenylacetaldehyde, which has an aroma reminiscent of flowers or honey:

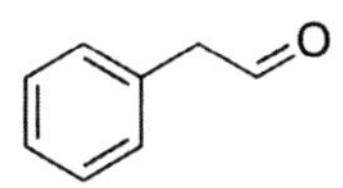

Aldehydes may have branched chains. The primary branched-chain aldehydes in cheese arc 2-methylpropanal:

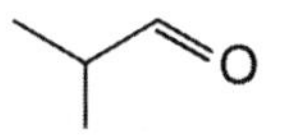

2-methylbutanal:

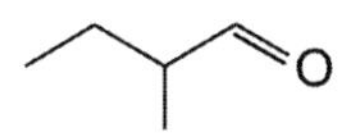

and 3-methylbutanal:

These three compounds appear to be responsible for nutty flavor in Cheddar, though by themselves they have a green or malty aroma.

TABLE 10.2

Factors affecting cheese texture

Factor	Effect on cheese texture	Reason
Moisture content is high	Softer	Rigidity decreases as amount of protein decreases
Fat content is high	Better stretching	Protein network is less dense
CCP content decreases	Weakens	Links between casein molecules decrease
pH is under 4.9	More brittle and grainy, less cohesive and stretchable	Casein molecules become more compact
pH is over 5.8	Crumbly, less cohesive and stretchable	Curd particles do not fuse together
Aging	Softens, becomes less stretchable and easier to fracture	Proteolysis breaks bonds between caseins

Texture is a sensory property, meaning that only people can describe it, not instruments. Texture has many characteristics deriving from the structure of the food, and is primarily detected by touch and pressure. Alina Szczesniak was a pioneer in defining textural attributes of food while at General Foods Corporation in the 1960s, and she classified texture descriptors into mechanical, geometrical, and compositional parameters. Table 10.1 shows these mechanical properties, adapted for cheese.

Fracturability (or brittleness) and chewiness are secondary properties that are included in "cohesiveness." "Geometrical properties" include particle size, shape, and orientation, and have descriptors such as gritty, grainy, and fibrous. "Compositional parameters" are moisture content (dry, moist) and fat content (oily, greasy). Texture changes as you eat a food because it is broken into small pieces and lubricated with saliva (Box 10.3). Cheese judges and graders also consider the size, shape, spacing, and number of eyes in Swiss-type cheeses to be a part of texture, though others refer to these aspects as part of the "body" of the cheese. We will talk about the ways to analyze texture in Chapter 15.

Several factors determine the texture of cheese by affecting the casein, shown in Table 10.2.

The pH at coagulation and draining affect the rate of acid production, which determines the mineral content of the cheese. For all intents and purposes, "mineral" means CCP, which increasingly dissolves in the whey as the pH goes down. For instance, the low pH of the interior of mold-ripened cheeses (4.6–4.7) causes much of the CCP to dissolve and migrate to the surface. Combined with proteolysis, these cheeses can become very soft on the inside, while the CCP collecting on the outside helps develop a rind. Stretching is particularly important with Mozzarella, and a number of studies have shown that pH has to be between 4.9 and 5.6 for the cheese to stretch acceptably. Washing of the curd also removes some calcium. We will look at washed curds next.

The woman on the phone had a voice that could melt the wax off a gouda.
—JUDITH STONE, *Light Elements*, 1991

11

WASHED CURD CHEESES, LACTONES, FEED, AND SPECIES

THE ABOVE QUOTE is the first sentence of a story about some of my work, in this instance distinguishing mislabeled cheeses from the real things. The rules and regulations about cheese will be covered in Chapter 16, but this chapter will explain why Gouda (and Edam) cheeses have wax in the first place, and why they, Colby, and Monterey Jack are in a class by themselves.

Gouda and Edam

Gouda comes in wheels and is made from whole milk, and Edam comes in spheres and is made from part-skim milk. Otherwise, the procedures for making these two classic Dutch-type cheeses are the same. The milk is cooled to 86°, starter is added, and rennet is added immediately. The curd is cut with 5/8-inch knives and stirred gently 20 minutes. Then the curd is scalded: a third of the whey is drained and replaced with water at 95° over a 15-minute period. This curd-washing step rinses out much of the whey and removes some lactose from the curd, resulting in a cheese with a sweeter and less sour flavor. Stirring continues until the temperature levels off at 100°, and then stopped. When the curd has settled, it is pressed for at least 10 minutes while still under the whey, using steel plates. The whey is drained, and the curd is removed, pressed, and brined for anywhere from a day (1-pound size) to a week (45 pounds). Cheeses intended for export are coated with red (or sometimes yellow) paraffin wax to prevent them from drying out; black wax is used if the cheese is to be aged a long time. Dutch-type cheeses were frequently used to feed sailors on long voyages and required a strong coating to stay in good condition. The cheese is initially stored for two weeks at 50°–55° and 75 percent relative humidity. The product is next moved to a room that is a few degrees warmer to encourage microbial growth. Both varieties develop some small eyes. After long aging, crystals of calcium lactate and tyrosine may form in Gouda, imparting some crunch. Industrial production of these varieties is always from pasteurized milk, but Boerenkaas ("farmer's cheese") is made on farms from raw cow's, goat's, or sheep's milk.

Gouda comes in wheels weighing anywhere from 1 to 88 pounds and Edam spheres weigh around 4 pounds. The cannonball shape of Edam once prompted its use as actual cannonballs (Box 11.1).

BOX 11.1
CANNONBALLS

"I was blown up while we were eating cheese."
—ERNEST HEMINGWAY, *A Farewell to Arms*, 1929

In the days of sail, iron cannonballs were fired from close range at enemy ships to damage their wooden hulls. Edam's spherical shape inspired Commodore John H. Coe of the navy of Montevideo, Uruguay, to use the cheese as actual cannonballs when they ran out of ammunition while battling Admiral William Brown's force from Buenos Aires in 1841. The Edam was rock-hard, and the first lieutenant said that he broke the carving knife trying to cut it. One cheese shattered into dangerous splinters upon striking the main mast, another flew in through a porthole and killed two men before fragmenting, and several others ripped through the sails. Brown retreated.

A 2009 episode of the *Mythbusters* television show attempted to show whether or not a cheese fired from a cannon at 30 feet (a typical distance in nineteenth-century sea battles) could put a hole in a sail. The Edam they used was too soft and simply bounced off the sail without damaging it. A smoked Gouda, which was hard and brittle, fragmented upon exiting the cannon barrel. They also tried a hard Spanish goat cheese, Garrotxa (gar-ROH-chah, named after a county), which did puncture the sail; its success was credited to its combination of firmness and flexibility. They did not attempt to splinter wood (or kill anyone), and it is possible that the Edam on Coe's ship was capable of wreaking the havoc reported.

Colby and Monterey Jack

Colby was first made in Colby Township, Wisconsin, in 1874, and the procedure for manufacturing it is similar to that of Cheddar, up to the whey-drainage step. The whey is removed until it is at the level of the curd, and 59° water is added until the temperature in the vat reaches 79°. Water is added, and the curd is stirred vigorously over a 15-minute period to prevent a tight structure from forming. The warmer the water added, the lower the moisture in the final product. NaCl is added after draining and the cheese is placed in hoops and pressed. The cheese is vacuum-packaged and aged at 45°–55° for up to three months.

Monterey Jack was developed by Franciscan monks in California in the seventeenth century and was pressed with a device called a "jack." It was popularized by businessman David Jacks in the late nineteenth century, who shipped it out of Monterey and took credit for its invention and name. This cheese differs from Colby in that all of the whey is drained, and the curd is allowed to sit with occasional stirring until the pH drops to 5.3. The cheese forms a rind before packaging and is aged for up to two months. Monterey Jack curds are sometimes mixed with Colby or Cheddar curds to produce Colby Jack or Cheddar Jack.

Havarti

Havarti is named after the farm north of Copenhagen where modern Danish cheesemaking was developed in the latter part of the nineteenth century. The cheese itself was introduced in the 1920s. It is similar to a mild Tilsit, but the curd is washed and the smear on the surface is not allowed to develop, making it an interior-ripened cheese like the other washed curd varieties.

Lactones

Lactones do not taste like cheese, but they are essential for the proper aroma and flavor of every cheese variety. Their concentration in milk fat is 10–30 ppm, which is well above their flavor threshold, making their fruity, sweet flavors easily detectable. All of the hydroxy acids in milk fat are converted to lactones upon heating, resulting in a lactone concentration over 80 ppm. That's why your kitchen is filled with such a lovely aroma when you bake foods containing butter. The most important lactones in cheese and how they are formed in Gouda are discussed in Box 11.2.

BOX 11.2
LACTONES

As mentioned in Box 5.4, lactones are formed from hydroxy acids, which are fatty acids that have an OH group attached to the carbon chain. A recent study on Gouda during ripening showed that the reaction takes place while the hydroxy acid is still attached to the triglyceride, and the lactone then splits off. The same mechanism should be true for other cheese varieties. Formation of lactones in cheese is spontaneous and does not involve bacteria or enzymes. Lactones in cheese may have as few as four carbon atoms and as many as eighteen, but the most important lactones have eight, ten, or twelve carbons.

γ-Octalactone (γ, gamma, is the third letter of the Greek alphabet) is similar to the δ-octalactone from Box 3.2 except that the ring contains four carbon atoms and the tail has four. Both lactones have a coconut or peach flavor. A similar compound with a CH_3 group attached to the ring is found in oak and in alcoholic beverages aged in oak barrels. It is often called "whisky lactone."

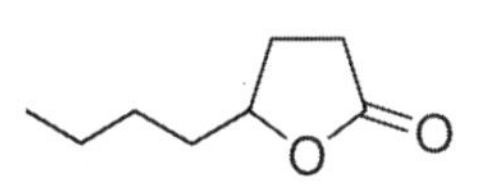

δ-Decalactone has ten carbons in all (*deca-* is the prefix for ten) and also imparts a peachy, coconut flavor. γ-Decalactone is responsible for apricot flavor.

δ-Dodecalactone has 12 carbons (*dodeca-* is the prefix for twelve) and has flowery and buttery flavors.

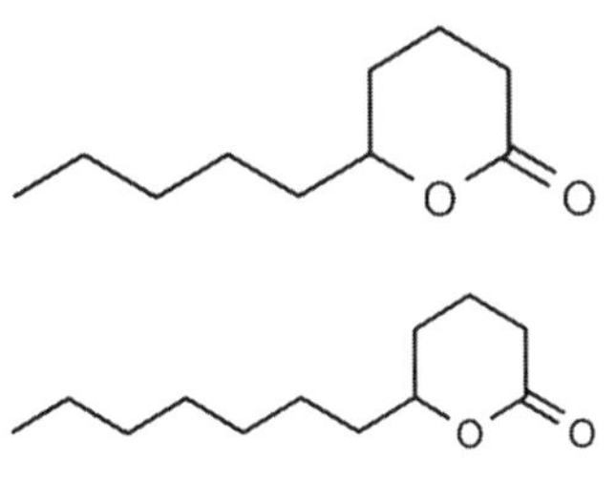

Fodder versus Forage

As mentioned in Chapter 3, most American cows are on fodder, which is conventional feed consisting of hay, silage, and total mixed ration. "Hay" is composed of dried legumes (especially alfalfa) or dried grasses (such as fescue and timothy). "Silage" contains a combination of alfalfa, barley, corn, and soybean that is harvested while still wet, chopped up, placed in a silo, and fermented for a few weeks by bacteria. "Total mixed ration" is a blend of various plants, protein supplements, vitamins, and minerals. A study of milk from cows living in similar farms in an area of Italy showed differences in the volatile compounds depending on the feed. When comparing three diets, milk from cows eating hay and corn was found to have higher levels of alcohols, aldehydes, and ketones than milk from cows that had silage in their feed (Box 11.3).

BOX 11.3
FEED

Not only does fodder result in different compounds in milk than forage does, but the type of fodder is important. The following table (Table B11.3) compares the number and concentration (measured in parts per billion) of volatile compounds in one study of milk from cows with different diets. All three feeds contained 9% soybean meal.

TABLE B11.3

Diet and volatile compounds in milk

Compounds		Feed		
Class	Number found	57% hay, 30% corn, 4% sugar beet and wheat bran	59% corn silage, 17% hay, 14% corn	15% grass silage, 51% corn silage, 11% hay, 14% corn
			PPB IN FEED	
Alcohols	8	51	24	7
Aldehydes	9	110	29	58
Ketones	8	213	124	184
Others	16	15	14	11
Total	41	389	191	260

The cows fed primarily with hay gave milk with higher levels of each compound class. Though tens and hundreds of parts per billion don't seem like much, remember that these numbers are for whole milk and that the resulting cheese would concentrate these compounds by roughly a factor of ten. Also, some of these compounds have low odor thresholds, making them noticeable.

The above study did not look at fatty acids, which are also passed on from the diet to the milk. In temperate climates, fresh grass contains 1%–3% fatty acids, with the higher number in spring and fall. Milk from cows on pasture has lower levels of myristic and palmitic acids, and higher levels of oleic, linoleic, and linolenic acids, than from cows on grass silage or corn silage. Table 2 shows the distribution according to a study in France (the silage tested also contained some concentrated feed).

BOX 11.3 (*continued*)

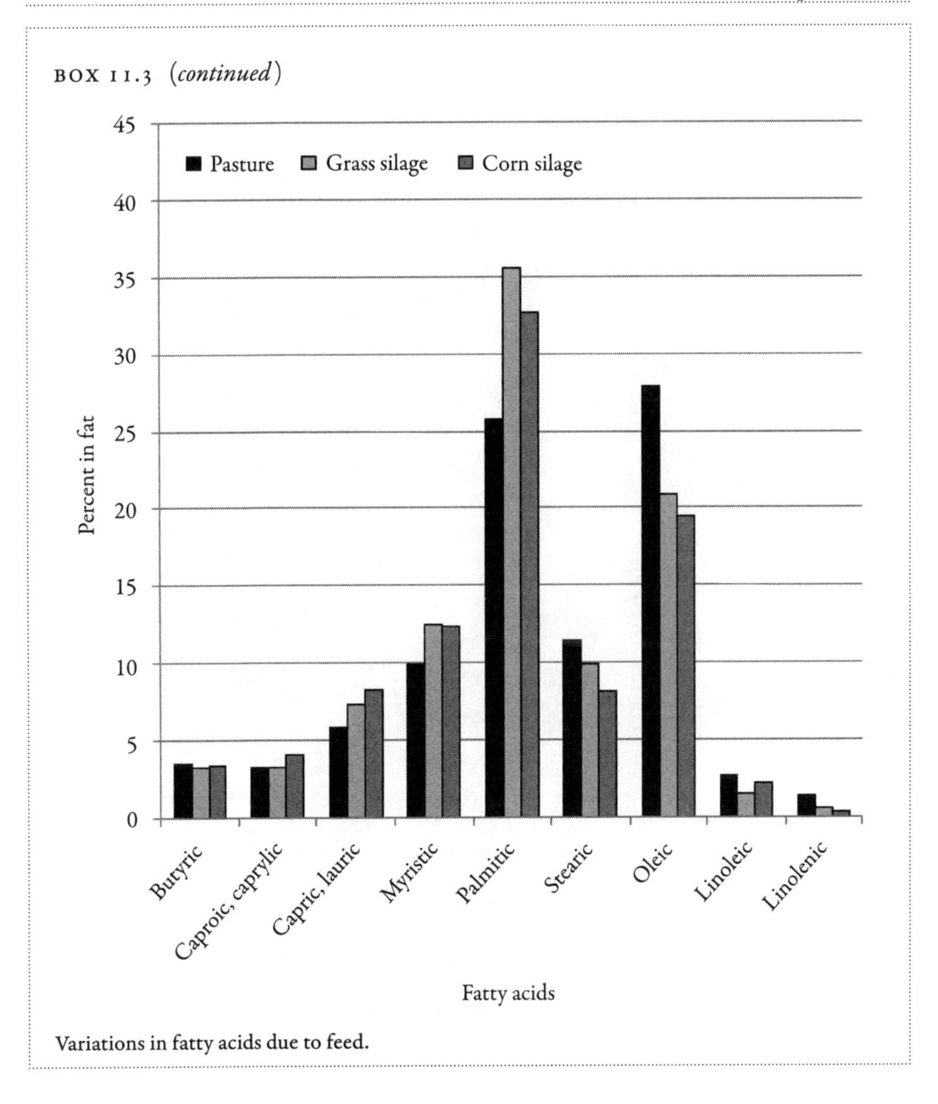

Variations in fatty acids due to feed.

Some cows and most sheep and goats are on pasture most of the year. Their milk will contain a larger variety of compounds than milk from animals on conventional rations. In another study in Italy, sheep were grazed on three different pastures: a cultivated pasture containing a mix of perennial ryegrass and clover; an uncultivated pasture containing grass and plants in the crucifer, knotweed, and pea families; and a pasture of oats. The plants from the uncultivated pasture contained 38 volatile compounds, compared with 16 in each of the other two pastures. Only nine compounds were common to all three. When the milk was made into caciotta (a generic,

FIGURE 11.1 Holstein and Jersey cows feeding silage.

BOX 11.4
DIFFERENCES BETWEEN SPECIES

The milk from cows, goats, sheep, and water buffalo is most often used to make cheese. Here are selected compounds from one study comparing volatiles in raw milk from these species:

TABLE B11.4

Volatiles in cheese from different species (in ppb)

		Species			
Compound	**Odor**	**Cow**	**Goat**	**Sheep**	**Water buffalo**
		ALCOHOLS			
Pentanol	Alcoholic	2	5	5	10
1-Octen-3-ol	Mushroom	0	8	4	10
2-Phenylethanol	Rose-like	0	0	0	10
p-Cresol	Stale	8	3	1	13
		ALDEHYDES			
Pentanal	Herbaceous	20	40	20	60
Heptanal	Herbaceous	40	10	4	10
Nonanal	Green	20	30	20	40
Decanal	Herbaceous	10	30	40	6
Phenylacetaldehyde	Floral	0	3	0	20
		ESTERS			
Ethyl butanoate	Apple	120	80	40	300
Ethyl hexanoate	Unripe fruit	90	70	40	150
Ethyl octanoate	Apple	80	50	70	50
Ethyl decanoate	Fruity	80	60	100	2
		KETONES			
2-Butanone	Varnish	1	1	0	13
2-Pentanone	Rotten	1	3	2	5
2-Hexanone	Blue cheese	1	3	1	2
2-Heptanone	Blue cheese	1	2	2	4
2-Nonanone	Ketone	1	2	2	4
2-Undecanone	Oily	1	3	1	3
		LACTONES			
δ-Decalactone	Creamy	0.8	1.0	0.5	0.3
δ-Dodecalactone	Coconut	0.9	0.5	0.4	0.2

(continued)

BOX 11.4 (*continued*)

Compound	Odor	Species			
		Cow	**Goat**	**Sheep**	**Water buffalo**
	NITROGEN-CONTAINING				
Indole	Stale	0.2	2	2	7
	SULFUR-CONTAINING				
Dimethylsulfone	Sulfurous	200	250	120	40

Water buffalo milk contained the highest concentrations of alcohols, aldehydes, and ketones, and the 1-octen-3-ol first mentioned in Chapter 6 was at the odor threshold in water buffalo milk only. The microbial population in the rumen of water buffalo is higher than those of the other species, allowing them to break down fatty acids and amino acids faster and to a greater extent. 2-Phenylethanol, for instance, is a degradation product of the amino acid phenylalanine (both are described in Box 5.1), and this reaction apparently does not take place as quickly in milk from cows, goats, and sheep.

The esters were the most important and potent odorants. Their concentrations in milk from cow and water buffalo decreased as the chain length increased; sheep's milk had the opposite trend. The lactones were present at low concentrations, but were detectable due to their low odor thresholds. Dimethylsulfone represented 20%–28% of the volatile compounds in cow's, goat's, and sheep's milks, but only 4% of the volatiles in water buffalo milk.

simply-made soft cheese) and aged for two months, the cheese made from the milk from the uncultivated pasture contained far more acetic acid and short-chained fatty acids than the other cheeses.

Pasture plants vary according to season. When caciotta was made from goat's milk, differences in the plants available depended on the time of year and resulted in differences in the cheese. Goats selected grasses as 85 percent of their diet in the winter, but new growth caused them to choose 60 percent legumes and herbaceous plants in the spring and 70 percent herbaceous plants in the summer. Winter and spring milk and cheeses had fruity and musty aromas due to alcohols, esters, and ketones. In summer, there were citrus, green, mint, and resinous odors from terpenes. We will discuss more about seasonality and terpenes in Chapter 13.

Species

In addition to food source, many volatile compounds that are found in milk and contribute to cheese flavor differ because of species. Cows, goats, sheep, and water buffalo metabolize food in different ways. A study demonstrated this by pooling the milk from about 100 animals of each of these species and determining the volatile compounds in them. All of the animals were pasture-fed in Caserta, Italy (the water buffalo), or Dijon, France (the others). The milk had dozens of compounds in common, but often in widely varying amounts (Box 11.4).

BOX 11.5
BOVINE SOMATOTROPIN

All animals produce growth hormone in their pituitary glands, where it helps regulate metabolic processes. In calves, bovine somatotropin (BST), also known as growth hormone (BGH), helps with growth by building tissue. In lactating cows, it allows body fat to be used less for growth and more for the energy necessary for milk production. Small amounts of BGH are passed on to milk, where 90% of it is destroyed by pasteurization.

In the 1930s, injection of cows with BST was found to prevent mammary cell death, causing milk yield to increase. The use of BST was not cost-efficient then since it could only be obtained by extracting it from bovine pituitary glands. But by the 1990s it became economically feasible to produce BST by genetically engineering *E. coli*. This recombinant BST (abbreviated rBST or rBGH) boosts milk production by 10%–25%, though it increases the possibility of mastitis, an udder inflammation. It was approved by the U.S. Food and Drug Administration (FDA) in 1993 after determining that its use would be safe to humans. The FDA found that BST was not biologically active when consumed by humans, and that the milk from cows treated with rBST was no different than milk from untreated cows. Furthermore, the FDA ruled that milk and milk products cannot be labeled as "hormone-free" since all milk contains hormones, including tiny amounts of BST. It is impossible to tell whether milk contains BST, rBST, or both. To date, over 30 million American cows have received injections of rBST.

Nevertheless, only the United States, Mexico, most countries in Central and South America, and several other nations allow rBST. Many dairies, cheesemakers, and other manufacturers proudly state on their labels that milk from rBST cows is not used in their products. Why is there so much fuss about rBST? For one thing, consumers are wary about any changes to their milk—they have a strong emotional bond with baby's first food. Many people are concerned with the health of cows on rBST. Also, large operations are more likely to use rBST than family farms, giving them an economic advantage that many view as unfair.

The flavor of cheese is important to consumers, but cheesemakers also have to look at yield, the amount of cheese they are able to make from a given quantity of milk. Species makes a difference here also. Back in Chapter 1 we saw that milk from sheep and water buffalo has more protein and fat than milk from cows and goats. When cheese is made, the milk is often standardized so that the ratio of casein to fat is 3 to 4. As a result, the yield of sheep cheese is 15 pounds per 100 pounds of milk, with water buffalo at 12 and cow and goat around 10.

How much milk do these animals provide? Dairy sheep average only 400–1,100 pounds of milk during a lactation period of 120–240 days, so their milk and the cheese made from it is scarcer and thus costlier than cow's milk cheese. Dairy goats average 1,200–2,600 pounds of milk during an average lactation period of 284 days. Only a couple of dairy water buffalo herds are in the United States, so cheese made from their milk is sold at a premium. These animals can be expected to produce around 2,600–3,300 pounds of milk during their lactation period of 200–230 days. Compare those numbers with the 15,000–24,000 pounds of milk per 305-day lactation for cows (see Box 11.5 for one reason for the high numbers). It is therefore no surprise that most cheese in the world comes from cow's milk, including all of the varieties discussed in the next chapter.

You are not approaching Nature in one of her myriad tints of mood, as in the holy act of eating cheese.

—G. K. CHESTERTON, *Alarms and Discursions*, 1911

12

CHEESES WITH EYES, FURANS, HYDROCARBONS, AND FOOD PAIRING

THE CHEESES IN this chapter are the "holy" (holey) Swiss types. These varieties originated in mountainous regions where the climate is cool and the lactic acid bacteria population is relatively low, resulting in slow acidification and whey expulsion. The NaCl levels are also low because of the cost of hauling salt up to the farms. The wheels are large due to lack of storage facilities for the milk—all of it is used to make one big cheese. These factors make reduction of moisture to the proper levels difficult, so the curds are cut into small pieces and heated above 120° to facilitate whey removal. The resulting cheese has a hard rind and a relatively elastic body.

The propionibacteria normally found in milk are able to grow best with low levels of acid and NaCl. These bacteria metabolize lactic acid to propionic acid (Box 12.1), and two of the compounds formed, calcium propionate and magnesium propionate, appear to be responsible for the sweetish flavor of Swiss-type cheeses. Propionibacteria also produce CO_2 gas, which forms the eyes within the supple structure of the cheese. In France, cheesemakers categorize the eye size as tiny (*les yeux de perdix*, "partridge eyes"), small (*petit pois*, "little peas"), medium (*cerises*, "cherries"), and large (*noix*, "walnuts"). Emmental has large eyes; Comté is classified as having small to medium eyes; and Tomme de Savoie has tiny eyes.

BOX 12.1

WHERE THE HOLES COME FROM

> *"On earth I was a manufacturer of Imported Holes for American Swiss Cheese, and I acknowledge that I supplied a superior article, which was in great demand."*
> —L. FRANK BAUM, *Dorothy and the Wizard in Oz*, 1908

The eyes in Swiss-type cheeses are created when *Propionibacterium freudenreichii* subspecies *shermanii* metabolizes lactic and propionic acid. The initial reaction looks like this:

$$3\ CH_3CH(OH)COOH \longrightarrow CH_3COOH + 2\ CH_3CH_2COOH + CO_2 + H_2O$$

3 lactic acid molecules form 1 acetic acid molecule, 2 propionic acid molecules, 1 carbon dioxide molecule, and 1 water molecule.

Extra oxygen and hydrogen atoms are also produced, and these go toward an enzyme that transports energy within the bacterial cells. The CO_2 gas accumulates at the weaker spots in the cheese matrix, forming the bubbles that we call "holes" or "eyes." The cheese is made with a relatively low NaCl level, which allows the bacteria to survive. Some 120 liters (32 gallons) of CO_2 are generated during ripening of a 175-pound wheel of Swiss cheese. About a third of the gas escapes into the air, half of it stays dissolved in the curd, and the rest forms eyes.

Swiss Cheese

Emmental (also called Emmentaler) is the name of the authentic Swiss cheese, originating in the thirteenth century in Switzerland's Emmen valley. The starter consists of the thermophile *Streptococcus thermophilus*, the thermophile *Lactobacillus delbrueckii* subspecies *lactis* (which metabolizes glucose and galactose, and governs pH development), and *Propionibacterium freudenreichii* subspecies *shermanii*, the adjunct responsible for eye formation through generation of CO_2. The *P. shermanii* grows slowly and at relatively high temperatures.

Emmental and similar cheeses are tricky to make because of the balancing act between achieving the proper texture and the generation of the characteristic eyes. If too much CO_2 is created, the cheese may explode, and if no CO_2 is made, the cheese will be "blind" (no eyes). The starter is added to milk at 98°, and rennet is added 15 minutes later. Quarter-inch knives are used until the curd is in small granules the size of rice grains. The curd is stirred for 30–60 minutes and heated to 125° over 30 minutes, with agitation to prevent the pieces from sticking to each other. The temperature is too high for bacterial growth, so the pH at this stage has only decreased to 6.3, which allows the CCP to be retained in the curd. Much of the chymosin activity is also halted at this temperature. After agitation is stopped, the curds are allowed to settle at bottom of vat under the whey. The curd is pressed with metal plates for up to half an hour before the whey is drained. Pressing continues for another 18 hours, during which the curd temperature drops and the bacteria begin to grow again and metabolize the lactose. The pH stabilizes around 5.2, which is ideal for curd elasticity (the cheese is stretchable when heated by the consumer) and allows the eyes to form. The curd is held in brine for two days and then stored at 75° for 6–8 weeks at 75 percent relative humidity. After a month, the *P. shermanii* starts to metabolize the lactic acid, creating bubbles of CO_2. The cheese is then stored at 50° as eye-formation slows. A firm, dry rind that prevents the escape of CO_2 is formed during storage. Eyes continue to be produced until refrigeration stops the process. The eyes are supposed to be up to an inch in diameter and 1–3 inches apart, with a 110-pound wheel containing a thousand eyes occupying a volume of 760 cubic inches. Often, eyes are not perfectly formed throughout a cheese; a list of possible defects is in Box 12.2.

Gruyère and Similar Cheeses

Gruyère (gree-AIR, named after a town in Switzerland) is made in the same basic way as Emmental but with smaller curds and longer, harder pressing. It is aged in cellars or caves at 96 percent humidity and a temperature of 56°. Gruyère has a few small, irregular eyes. Graviera (Greek for Gruyère) is similar but may be made from goat's or sheep's milk. Comté (French for "county" and made in Franche-Comté and the Rhône Alps) is also called Gruyère de Comté and is the most-produced cheese in France. It is made mostly from Montbéliarde cows, and it takes 120 gallons to make an 80-pound wheel. The milk must not travel more than 15 miles to the cheese plant, and it is heated only during coagulation. Beaufort (bow-FORE, named after a French town) is similar to Gruyère but uses only *L. helveticus* as a starter. A wheel has inwardly curving sides because it is encircled by a beechwood belt during aging. The cloth rind is washed with brine containing whey and older cheese. Beaufort cheeses made in winter are white because of the hay fed to the cows; in summer the cheeses are light yellow because the cows are on pasture.

BOX 12.2
EYE DEFECTS

The eyes in Swiss-type cheeses should be around a quarter-inch in diameter, well-distributed, with a glossy, velvety appearance on the interior. The USDA changed their regulations in 2002, allowing smaller eye size because the slicing machinery would jam up with the larger eyes. The common eye defects come under five categories and are shown here.

I'm not sure where "streuble" comes from, but "cabbage" is a descriptor because the multitude of eyes results in a thin layer of cheese between them, giving the cheese a cabbage-like appearance. Checks, cracks, and splits are a result of late fermentation (when propionic acid breakdown restarts late in the ripening process) or when butyric acid is fermented. The body of the cheese cannot handle the additional gas generated, and begins to fracture. Eyes with the shell defect resemble a nutshell.

TABLE B12.2

Defects in Swiss-type cheeses

Defect	Description	Likely cause
	DISTRIBUTION:	
Gassy	Eyes of various appearance, shape, and size	Unwanted microbial growth
Nesty	Many small eyes in a small area	Improper pressing
Streuble	Many small eyes just under surface	Curd temperature differences
Uneven	Eyes unevenly distributed	Curd mishandling
One-sided	Eyes well-spaced on one side, lacking on other	Curd temperature differences
	NUMBER:	
Blind	No eyes	Lack of microbial activity
Underset	Few eyes	Too little microbial activity
Overset	Excessive eyes	Unwanted microbial growth
	SIZE:	
Blowhole	Air bubble at least 5 inches in diameter	Poor pressing

BOX 12.2 (*continued*)

Defect	Description	Likely cause
Large eyes	Eyes between 13/16 and 1 inch in diameter	High CO_2 generation
Small eyes	Eyes between 1/8 and 3/8 inch in diameter	Low CO_2 generation
Afterset	Small eyes caused by unwanted late fermentation	Poor quality milk
	SHAPE/TEXTURE:	
Cabbage	Cheese overcrowded with many irregular eyes	High moisture and soft body
Collapsed	Flattened eyes	Body too soft
Checks	Small, short cracks in cheese	Excessive proteolysis
Picks	Small irregular openings in cheese	Excessive proteolysis
Splits	Sizable cracks	Excessive proteolysis
Frogmouth	Eyes have spindle shapes	Curd too acidic and too firm
Irregular	Eyes are distorted or elongated	Moisture within block varies
	INTERIOR:	
Dull eyes	Eyes lack some glossy luster	Insufficient whey
Dead eyes	Eyes lack glossy luster	drainage, leading to
Rough	Eye walls lack smooth appearance	high moisture, excessive acidity, and
Shell	Multi-dimensional wall appearance	too firm body

Tomme de Savoie

Tomme de Savoie ("alpine cheese of Savoy") is usually made from milk left over from the manufacture of other varieties such as Beaufort. The differences between producers and seasons are noticeable with this cheese, which is made from partly skimmed milk. The curd is uncooked and pressed into 3-pound cylinders. The cheese is often seasoned with cumin.

Other Cheeses with Eyes

Samsø (named after a Danish island) is a softer version of Emmental, with a milder flavor. It comes in blocks or wheels weighing 31 pounds. Jarlsberg (named after a Norwegian county) has many similarities to Emmental, including the use of *P. freudenreichii* subspecies *shermanii*, but its manufacturing procedure has been secret since production began in the early 1960s. The wheels are 20–24 pounds. Appenzeller (named after a Swiss town) is unusual in that the rind is soaked in brine containing a mixture of herbs, spices, and cider or wine. It may be brushed or submerged for a month or two, imparting a unique flavor to the cheese.

Furans and Hydrocarbons

Some strains of Lactobacilli generate furans, in particular furaneol, homofuraneol, and sotolon (Box 12.3). Furans are formed in a wide variety of foods during breakdown of carbohydrates, and presumably arise from lactose degradation in cheese. Furaneol and

BOX 12.3
FURANS

Furans are compounds containing one oxygen atom and four carbon atoms in a ring. Two furans that affect cheese flavor are furaneol and homofuraneol, both of which have a strawberry or caramel aroma. The structures are shown here with their common names and full names. This is furaneol, or 4-hydroxy-2,5-dimethyl-3(2*H*)-furanone:

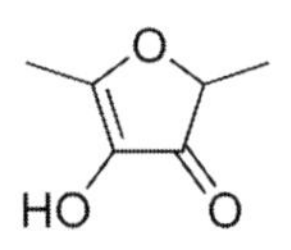

You can see why chemists often prefer the common names. Here is homofuraneol, or 5-ethyl-4-hydroxy-2-methyl-3(2*H*)-furanone:

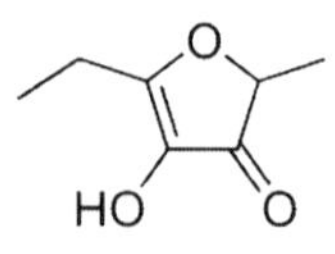

Another furan found in cheese is sotolon, or 3-hydroxy-4,5-dimethylfuran-2(5*H*)-one, which has a maple syrup aroma:

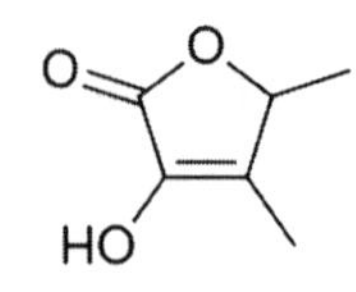

homofuraneol, which impart a caramel aroma in cheese, have been identified as having a high impact on Emmental flavor; their concentrations in ripe Emmental are higher than those of any other flavor compound except δ-decalactone and 2-heptanone. Furaneol, homofuraneol, and sotolon are also partly responsible for Cheddar aroma.

As you might guess from the name, hydrocarbons are molecules that contain just hydrogen and carbon. Cheese contains relatively low levels of hydrocarbons, which are breakdown products of fatty acids and chlorophyll from the plants that the animals eat. Some are described in Box 12.4. Most hydrocarbons in cheese do not contribute to flavor since they are relatively big molecules and are not volatile.

Food Pairing

Cheese is the food that most resembles wine: both preserve a liquid food, both are matured by the action of microorganisms, and both travel well, improve with age, and have many varieties that cannot be exactly duplicated at a different location.

BOX 12.4
HYDROCARBONS

A volatile hydrocarbon found in cheese is toluene, which smells like paint thinner and fortunately is not present in large amounts. The compound resembles the phenol we first saw in Box 6.1, but the OH group is replaced by a CH_3.

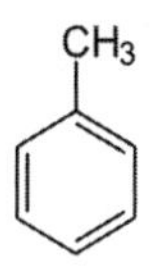

The larger hydrocarbons, containing 12 or more carbon atoms, are not volatile and therefore impart no aroma to cheese. Smaller hydrocarbons do carry odors, though they are usually unpleasant ones. The 2-hexene shown below has a smell reminiscent of petroleum.

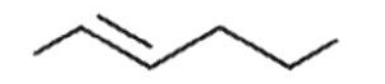

Phytene is found in chlorophyll and vitamins E and K. It has 16 carbon atoms in its "backbone" and another four attached at carbons 2, 6, 10, and 14. The double bond can be after the second carbon, as seen below, or after the first.

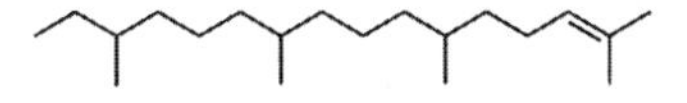

Squalene is produced by all plants and animals, and is used in humans to make cholesterol, steroids, and vitamin D. This behemoth contains 30 carbons and 50 hydrogens. Squalene is a terpene, which we will cover in the next chapter.

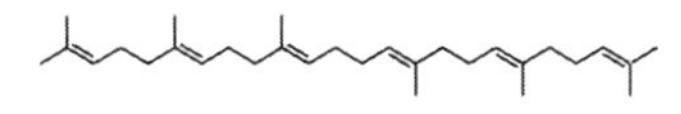

Unlike wine, cheese provides a wide range of textures that add to the enjoyment of consuming it. It is no surprise that wine and cheese go together. Beer is another fermented beverage that is traditionally consumed with some cheeses, especially British ones like Cheddar and very strong ones like Limburger.

Why do some foods (we're including beverages as foods) "go together" and others don't? One of the reasons is cultural. In Mexico, most restaurants have small limes on each table so the diners can squeeze the juice on their food, but this practice is rare north of the border. Macaroni and many cheese varieties were invented in Italy, but most Italians never eat macaroni and cheese together (the combination originated in America). A recent study showed that the more flavor compounds that are shared by two ingredients, the more likely they are to appear in North American and Western European recipes and the less likely they are to appear in East Asian and Southern European recipes. People tend to eat foods in combinations they are accustomed to consuming.

Another reason some foods are eaten together may be the presence of similar compounds in both. Foods with flavor compounds in common are often complementary as long as they are not overpowering (Box 12.5). This flavor-pairing theory is controversial and has not been proven. If you accept the theory, then you can say that wine and beer go with cheese because they contain similar products of microbial breakdown. In winemaking, grape juice is initially fermented with yeast in open vats, and then undergoes secondary fermentation in sealed stainless steel vessels or oak barrels, where lactic acid bacteria metabolize the malic acid that naturally occurs in the grapes. In addition to adding Lactobacillus and Pediococcus species, which produce lactic acid and its breakdown products, winemakers include *Oenococcus*

BOX 12.5
FLAVOR PAIRING

"But I, when I undress me
Each night upon my knees
Will ask the Lord to bless me
With apple-pie and cheese!"
—EUGENE FIELD, *The Poems of Eugene Field*, 1919

Most people do not enjoy eating two foods together if they have the same basic taste, because sensitivity to that taste is diminished and less desirable tastes become

BOX 12.5 (*continued*)

pronounced. Sweet foods and red wine both contain sugar; when consumed together, the tannins become more noticeable as one becomes desensitized to the sweetness.

On the other hand, people might like consuming two foods at the same meal if they have smaller amounts of flavorful common compounds. Wine, beer, and cheese contain acetaldehyde, phenylethanol, esters such as ethyl acetate, and some sulfur-containing compounds. Mozzarella, Parmesan, mushrooms, and tomatoes have compounds in common, including 4-methylpentanoic acid, which could help explain why pizza topped with mushrooms is popular. This is the basis for pairing some foods and wines with cheese. François Chartier, a chef and sommelier in Quebec, has explored the compounds that food and drink have in common. In his *Taste Buds and Molecules*, he lists a number of foods that should go with cheese because they contain compounds such as acetoin, linalool (described in the next chapter), and sotolon. Blue-veined cheeses and chocolate have over 70 compounds in common, and a few restaurants are preparing dishes featuring both.

But not everyone agrees with this idea. Some other compounds amplify or reduce flavors and may account for perceived pairing of flavors. For example, glutamic acid, a noted flavor-enhancer, is also found in the pizza components listed above, which may be the reason people like to consume them in one dish. Commenting on flavor pairing, Professor Hildegarde Heymann of the University of California–Davis told me "I personally do not quite buy it. I think ideal pairings have a great deal of situational effects that are not very dependent (or large) in terms of actual sensory effects." She and graduate student Berenice Madrigal-Galan used a trained panel to taste eight red wines (two each of Cabernet Sauvignon, Merlot, Pinot Noir, and Syrah) and eight cheeses (Gorgonzola, Stilton, Emmental, Gruyère, two Cheddars, Mozzarella, and Teleme) to see if wine flavor was affected by cheese, and vice-versa. They showed that, in general, each individual cheese had the same effect on all eight wines, finding that most of the wine aromas and flavors were suppressed. They theorized that binding of casein with volatile wine components prevented some of the aroma from being detected. The formation of a coating of milkfat in the mouth may also have prevented volatilization while producing a physical barrier to aroma compounds. Sourness in the wines was suppressed because the salt in the cheese counteracted the acid in the wine. The butter aroma in wine, originating in the diacetyl, was the only attribute that was enhanced. The authors concluded that any preferred cheese could be enjoyed with any preferred red wine.

Some scientists feel that flavor pairing is a fad, that we have a shortage of data, and that we know too little about the workings of the brain in processing taste and aroma inputs from various receptors. Others feel that the concept makes sense and that a real effect exists, even if it can't be explained yet. At some point, we will find out which side is right.

oeni, which generates diacetyl. Some wines (and, as we have seen, some cheeses) have a distinctly buttery flavor from the diacetyl. In beer brewing, malted barley is mixed with hot water, converting the starch to sugar. The sweet liquid is boiled, hops are added for flavor, and yeast is added after cooling to ferment the beer. Cheese and beer originate with grass and grain, and one may argue that they are closer relatives than cheese and wine, which starts with grapes.

A specific reason for pairing cheese with wine or beer is the presence of tannin, a compound that binds to amino acids and proteins and is found in some plants. Tannin from oak bark is sometimes used to tan animal hides into leather, hence its name. Tannin is formed in grape skins, seeds, and stems and is found in red wine, which comes from the pulp of red or black grapes that are fermented with the skins. White wine, which contains little tannin, is made by fermenting juice from which the skins have been removed. Tannin is also found in hops and malt, and is desirable in some beers and undesirable in others (in lager it can combine with protein and form a haze). A disadvantage of tannin is its astringency, resulting in a puckering mouth and a bitter aftertaste. Food with high levels of fat and protein, such as cheese, help neutralize the effects of tannin on the mouth. Milk also works, explaining why some people add it to tea, which also contains tannins. The salt in cheese also helps inhibit bitterness, which is found in some beers.

Wine and Cheese

Food and restaurant consultant Clark Wolf agrees with the University of California–Davis study noted in Box 12.5, saying that most cheese goes with whatever wine is left from the meal, and that one should not struggle with having to choose a wine to match a cheese. He prefers serving cheese with ripe apples and pears, dried fruit, nuts, and flatbread. He does admire the people who go to great lengths to pair cheese with just the right wine or beer.

If you are inclined to pair wine and cheese, it is probably best to consider it more an art than a science. Consuming flavorful food begins with the "attack," or initial sensory impressions, and ends with the "finish," or aftertaste, which involves the retronasal aroma; everything in between is savored when drinking wine and eating cheese. Paula Lambert of the Mozzarella Company in Dallas emphasizes balance and harmony, meaning that intensities should be similar: delicate cheeses with lighter wines, strong-flavored cheeses with young robust red wines, strong pungent cheeses with young full-bodied or sweet dessert wines, and aged mellow cheeses with older robust wines. Berkeley, California, food writer Laura Werlin lists ten rules, including the pairing of textures, where a light wine would go with a light cheese. She also suggests pairing opposites (such as a sweet wine with a salty cheese) and using white wine, which has high acid and buttery and fruity notes and may go with some fresh cheeses. *Maître Fromager* (certified cheese expert) Max McCalman in New York suggests matching fullness and persistence of flavors, complexity, aromas, appearance, and balance (sugar/salt, acidity, fat in cheese, and tannin in wine). He also considers *terroir* because, if the wine and cheese come from the same area, the water and soil will contribute to flavors of both ("if they grow together they go together"). A further discussion on *terroir* is in the next chapter.

In my snuff box I keep a piece of Parmesan
cheese—a cheese made in Italy, very nutritious.
—ROBERT LOUIS STEVENSON, *Treasure Island*, 1883

13

VERY HARD CHEESES, TERPENES, AND *TERROIR*

PARMESAN HAS LONG been highly regarded. In the 1350s, Giovanni Boccacio wrote of a place featuring a fountain of grated Parmesan in his allegory *The Decameron*. As the Great Fire of London approached their homes in 1666, British naval administrator Samuel Pepys and Sir William Penn (father of Pennsylvania's founder) dug a hole and buried some valuable possessions: wine and "Parmezan." Thomas Jefferson watched Parmesan being made on a trip to Rozzano, Italy, in 1787, detailed the process in his journal, and made sure it was always served at the White House when he became president. A northern Italian bank maintains a vault of 300,000 wheels of Parmesan, worth $200 million, used as collateral for loans. Some consider it to be the king of cheeses.

All of the very hard cheeses have thick rinds on the exterior and low moisture in the interior, allowing for long-term storage without refrigeration. A multitude of flavor compounds evolve during the lengthy aging period.

Parmigiano-Reggiano and Parmesan

Parmigiano-Reggiano is a *grana* ("grainy") cheese composed of fine, brittle granules. It originated in the thirteenth century in the Italian provinces of Parma and Reggio Emilia; if the cheese is made outside of that region, it is called Parmesan. The cheese is made in copper-plated steel kettles that are much wider at the top than the bottom. Milk from evening milking is allowed to form a cream layer, which is skimmed off; the remainder is mixed with whole morning milk. Whey from the previous day is incubated to serve as a starter culture in Parmigiano-Reggiano, and commercial starters are used for Parmesan. Lipase powder (for Parmesan) and calf rennet (for both) are added after 15–30 minutes at 108°, and the curd is cooked at 130° over 25 minutes. Rapid stirring of the curd (so the pieces do not stick to each other), high heat, and relatively high pH level (6.2–6.3) result in extensive syneresis. The curd is turned over and broken up into pieces the size of wheat grains using a sharp-edged wire whisk known as a *spino* or "thorn-bush." The curd collects in the bottom of the kettle for 30 minutes and is moved into a large hemp cloth bag, which is lifted by two people and slung from a wooden beam placed across the top of the kettle. After being divided into two masses and draining for a day, the curd is transferred into round hoops. The cheese is salted in brine or three weeks and ripened at 60°–64° and 85 percent relative humidity for up to two years. The cheese is graded after 14 months by striking it with a hammer and listening to reverberations. Parmesan is typically found in wheels weighing 55–85 pounds. Grana Padano, which comes from the Padano (Po River) area, is made in a similar manner, using partially skimmed milk from one milking.

Asiago

Asiago comes from the Asiago plateau (elevation 3300 feet) in Northern Italy and comes in *Pressado* ("pressed") and *d'Allevo* ("raised," referring to the mountainous area) versions. Pressado is made from whole cow's milk heated to 95°. The starter consists of whey from the previous run, and a rennet paste is added. The curd is cut and cooked at 113°. The curd is dry-salted, transferred into wooden hoops, and pressed for four hours. The wheels are wrapped with plastic bands and held 2–3 days at 50°–59°. The bands are removed and the wheels are placed in brine for two days. Finally, the cheese is aged at least three weeks. Asiago d'Allevo is made from partly skimmed milk that is cooked at 104° and again at 117°. Pressing is performed by stacking several cheeses atop each other and turning them several times. Asiago *stravecchio* ("very old") is aged for 15 months and is very hard. Asiago *mezzano* ("middle") and Asiago *vecchio* ("old") are aged at least 3 and 10 months, respectively. Asiago comes in wheels weighing 24–33 pounds.

Other Very Hard Cheeses

Romano is traditionally made from whole sheep's milk and called Pecorino Romano; it may also be made from cow's milk (Vacchino) or goat's milk (Caprino). The starter may be commercially supplied or from the previous day's whey. The curd is coagulated at 100° using rennet paste, cooked at 113°–115°, and cut into 1/8-inch granules, the size of rice grains. Romano is aged for eight months. The whey from Romano manufacture is often used to make Ricotta. Many other Pecorino cheeses are made elsewhere in Italy and are usually similar to Romano.

Roncal is an example of milk obtained from *transhumance*, the seasonal movement of herders and their flock between lowlands in the winter and highlands in the summer. Roncal is a hard and granular raw sheep's milk cheese made in northern Spain since the thirteenth century. It is acidified by the naturally occurring microflora, and cooked around 100°. After the whey is slowly drained, the curds are pressed against the sides of the vat, placed in forms, and pressed normally. It is smoked and ripened at 45° and 100 percent humidity for seven weeks. The Spanish island of Menorca is the home of Mahón-Menorca, a cheese that is tied up in cloth and pressed by pushing down on the knot with the hands, imparting a cushion shape. The cheese purchased after a few weeks of aging and ripened by *afinadores* (Spanish for *affineur*). After 10 months it has a texture similar to Parmesan's and a flavor with notes of peaches and sea salt.

Bitto, named after an Italian river, is made with cow's milk with up to 10 percent goat's milk allowed for added flavor. It is produced in Sondrio and Bergamo provinces from June through September. In another example of transhumance, the herds move from lower to higher elevations in those months and then move back down to the previous pasture, where new plants have grown. The cheese is made in copper kettles, transferred to wooden forms, dry-salted for three weeks, and initially aged in mountain huts. The remainder of the aging, which takes a year or two, and sometimes 10 years, is performed in factories in the valley.

Kefalotyri (keh-feh-low-TEE-ree, Greek for "head cheese," referring to the shape) is made from goat's or sheep's milk standardized to 6 percent fat. The curds are cooked at 111° and subjected to light pressure that is slowly increased. The next day the cheeses are both brined and dry-salted, and are also washed with a cloth soaked in brine. Not surprisingly, Kefalotyri has a strong flavor of salt.

Schabziger or Sapsago (sop-SAH-go, a corruption of Schabziger and also spelled Sap Sago), was originated by Swiss monks in the eighth century and has required a specific procedure for manufacture and stamp of origin since the fifteenth century. Nonfat milk is coagulated, the whey proteins are also coagulated, and the mixture is pressed and ripened for six weeks until dry. The fat-free cheese is then ground, salted, and mixed with the fenugreek herb. The cheese is fashioned into truncated cones, pressed, and ripened until hard enough to be used as a grating cheese (*Schab* is German for "scrape" and *Ziger* is the name for the curd produced in this manner). This "novelty cheese" is grayish green.

Our ability to store hard cheeses for a long time allows us to make them for special occasions and keep them around for a while, as described in Box 13.1. Suffolk cheese could also be held for an extended period, but it is hard to imagine an event in which someone would want to eat it (Box 13.2).

BOX 13.1
BABY CHEESE

Sbrinz (from Brienz, a town) and Saanen (another town) are very hard cheeses from the Swiss Alps. Sbrinz is similar to (and may be inspiration for) Parmesan, and is purported to be the oldest European cheese. Some wheels of Saanen are literally the oldest cheeses in Europe—they are often made to commemorate a birth, and pieces are eaten on special occasions throughout the person's life. If he or she has a notable career, small pieces of the cheese may be consumed for decades after his or her death. Sbrinz and Saanen are full-fat cow's milk cheeses cooked at 130°. Sbrinz weighs 80 pounds, Saanen weighs 12–25 pounds, and both are aged at least two years.

In England there was a superstition, lasting through the nineteenth century, in which a husband would have a wheel of hard cheese made to insure his wife's good delivery. Called a "groaning cheese," alluding to her complaints during childbirth, it was eaten from the center outward, starting when the baby arrived and continuing until the cheese became ring-shaped. The newborn was passed through the middle of it on the day of the christening. Small pieces of the first cut of the cheese were placed in the midwife's smock to cause young women to dream of their lovers, or placed under the pillows of young people for the same reason.

BOX 13.2
SUFFOLK CHEESE

> *"And, like the oaken shelf whereon 'tis laid,*
> *Mocks the weak efforts of the bending blade;*
> *Or in the hog-trough rests in perfect spite,*
> *Too big to swallow and too hard to bite."*
>
> —ROBERT BLOOMFIELD, "The Farmer's Boy: A Rural Poem," 1800

Bloomfield was writing about Suffolk cheese, which was made in that English county. It was manufactured from full-fat milk until 1650, when cattle disease reduced milk production. To recoup their losses, dairies began to skim off the fat from their milk to make butter, which sold at a premium in London, and made Suffolk cheese from skim milk. Many consumers griped about the hardness of the cheese: a 1661 diary entry by Samuel Pepys mentioned that he and his wife were "vexed" at the servants for grumbling about having to eat Suffolk cheese. But it was cheaper than cheeses with normal fat contents, and in 1677, Pepys himself drew up standards requiring Royal Navy sailors to receive 4 ounces of Suffolk cheese or 2.7 ounces of Cheddar three days a week. The basic cheese ration remained in effect for another 170 years, but Suffolk was replaced in

(*continued*)

BOX 13.2 (*continued*)

1758 when cheesemakers began to reduce the aging time in response to heavier demand during the Seven Years' War, resulting in even more complaints. By then, Suffolk was known as a very hard cheese that became even harder with age (sailors were said to have carved it into buttons), but the manufacturing shortcuts taken could cause it to be too soft, which led to spoilage and infestation with long red worms.

Stories about the hardness of Suffolk abounded. In the 1720s, Daniel Defoe referred to Suffolk County as being "famous for the best butter, and perhaps the worst cheese, in England." Bloomfield referred to it as "the well-known butt to many a flinty joke." In 1885, Charles Dickens noted the saying that "hunger will break through stone walls or anything but Suffolk cheese" and related the tale of Suffolk bound for India packed in sheet tin—the rats gnawed through the tin but could not eat the cheese. Suffolk cheese became scarce by the 1950s, though new versions are now being made artisanally.

Terpenes

Terpenes are found in many plant species and are thought to serve as a defense against microorganisms. Over 60,000 terpenes have been identified, and some of them are volatile, pleasant-smelling compounds found in milk from animals that have been allowed to graze. Box 13.3 shows a few of these compounds; Asiago, for instance, contains other terpenes not on the list, with the composition depending on the plants the animal has eaten. Alpine cheeses have considerably more terpenes than cheeses from the lowlands, and scientists can tell if an Asiago sample is the genuine article from the Asiago plateau by examining the terpenes in it. This leads us back to *terroir*, first mentioned in Chapter 3.

More on *Terroir*

Large-scale commercially made cheese is manufactured to be basically the same throughout the year. The product of a Cheddar factory in California in December will not vary appreciably from the same company's Cheddar made in Wisconsin in June. In contrast, a piece of farmhouse cheese is like a piece of geography. Economics (Box 13.4) and local know-how also come into play in the making of a particular variety in a particular location. But *terroir* mostly consists of a number of natural factors that interact with each other to create a unique product:

Vegetation is an obvious contributor to cheese flavor, since flavor compounds vary with plant species. The Sicilian variety Ragusano was found to contain 27 odor-active compounds after four months of aging when the milk was obtained from cows on pasture; only 13 were detected in cheese from cows on conventional feed.

BOX 13.3
TERPENES

Terpenes are absorbed by animals grazing on pasture plants and are transferred to their milk. Their levels in cheese do not change appreciably during aging. Terpenes are commonly found in tree resin (the word comes from *turpentine*), but are also formed in flowers and spices. These compounds are derived from isoprene, which has this structure:

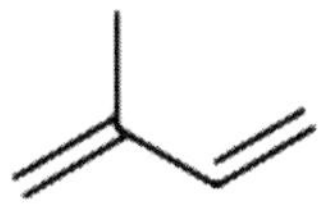

Isoprene (short for isoterpene) was first isolated from natural rubber and is found in many naturally occurring compounds, including camphor, menthol, and vitamins A and E. Here are some of the terpenes that have been found in cheese:

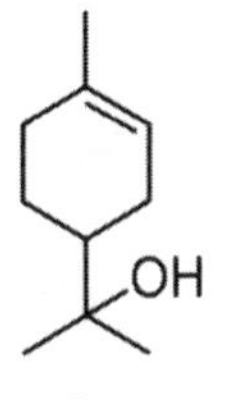

Terpineol, which has a lilac aroma, comes in α, β, and γ forms, depending on the position of the double bond. α-Terpineol is shown at right.

Limonene smells like citrus fruit and is found in citrus rinds (the chemical name comes from *lemon*). As mentioned in Chapter 5, many molecules come in left-handed and right-handed forms. Limonene is one of these. The right-handed form is R-limonene, which has an orange aroma. R comes from *rectus,* Latin for right. The thicker vertical line at the lower part of the molecule means that we can imagine it coming toward us.

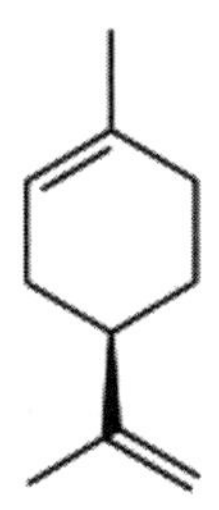

The left-handed form is *S-limonene,* which smells more like lemons. S comes from the Latin *sinister,* left. The dashed lines mean that the lower section is pointed away from us.

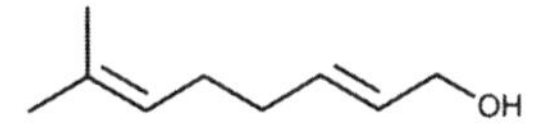

(*continued*)

BOX 13.3 *(continued)*

Another terpene is *linalool*, which has a floral flavor, though it is also used as an insecticide for cockroaches and fleas.

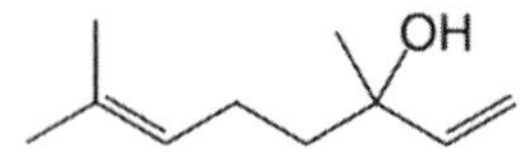

α-Pinene is responsible for the characteristic aroma of pine trees. It is a bicyclic compound, meaning that it has two rings. The rings are at angles to each other, which is why the structure is represented in three dimensions. Again, the thicker lines can be visualized as sticking out toward us.

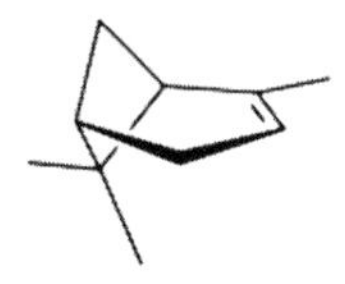

Geraniol and the similar compound nerol impart rose aromas. Terpineol, limonene, linalool, geraniol, and nerol are all found in grapes and hops and are transferred into wine and beer.

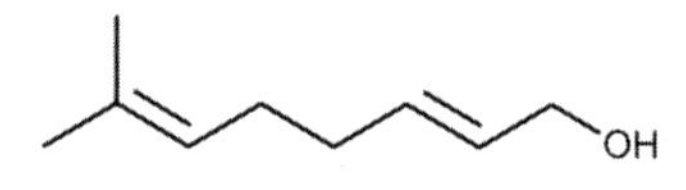

2-Methylisoborneol has a musty smell that contributes to the aroma of Brie and Camembert, as well as the scent of moist soil. It is also responsible for an off-flavor and bad odor of drinking water, and scientists have been trying to determine how to inhibit its formation in water sources. Two enzymes make it from geraniol diphosphate (geraniol with a pair of PO_4 groups attached to the oxygen atom), converting it into a bicyclic compound.

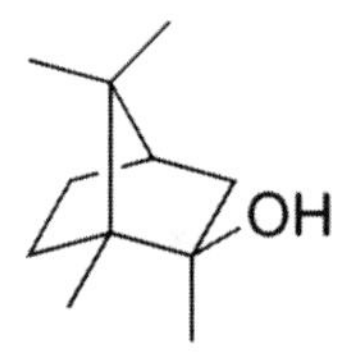

Climate is measured by rainfall, temperature, and the amount of sunlight reaching the land. It determines the vegetation that is able to grow in a particular region and affects the breeds of animals that can flourish there.

Animals are selected by the farmer to match the climate and vegetation. Cows prefer lush pasture, whereas goats and sheep are able to deal with more arid and less verdant areas. Moreover, we saw in Chapter 1 that fat and protein content vary with cow breed, which affects the properties of the cheese.

BOX 13.4
LOCATION, LOCATION, LOCATION

"A poet's hope: to be, like some valley cheese, local, but prized elsewhere."
—W. H. AUDEN in *Collected Poems*, 1976

Why are certain varieties of cheese made in specific places? A major reason is tradition, but another has to do with the economics of *terroir*. One aspect of the economics in the United States is the modern back-to-the-land movement, which has provided the incentive for dairy farmers to produce artisanal cheeses with a "sense of place." A farmer/cheesemaker often sets out to identify a variety that could be produced on site, and if successful, markets it with emphasis on its handmade origin and unique geography. Many consumers appreciate the land-stewardship that farmhouse cheeses imply and consider these products to be preferable to mass-produced ones.

Another economic facet is the ability to sell the cheese in a far-off market. The softer, more perishable cheeses were traditionally made to be consumed in the immediate area, while the harder, longer-lasting ones (such as Edam) were frequently manufactured with long-distance transportation in mind. Cheesemakers experiment with different varieties until they find one that is not just special, but also profitable. Before refrigeration and automobiles, Vermonters came to realize that a sturdy Cheddar would survive the trip to the lucrative New York market better than a high-moisture cheese that aged quickly. Cows are more viable in the Vermont landscape than goats or sheep, so the cheese was made with cow's milk. Vermont specializes in cow's milk Cheddar to this day.

Season of the year affects milk composition and therefore the cheese made from it. For example, research in our laboratory revealed seasonal variability in Queso Chihuahua, a cheese based on Cheddar and traditionally made from raw cow's milk by Mennonite communities in Chihuahua, Mexico. Their average September rainfall is 3 inches, but they receive only two-thirds of an inch per month in December, January, and May. The average temperatures are 46° in December and January, and 64° in May and September. These differences affect the feed and consequently the fat content in the cheese, which was 35 percent in the summer and 31–32 percent in the winter and spring. As a result, winter cheeses were significantly harder, springier, stiffer, and less cohesive than summer cheeses, with May cheeses in the middle.

Altitude affects the types of plants growing in pastures. Research on Abondance, a French cow's milk variety similar to Gruyère, found that it undergoes more protein breakdown and is less elastic when the milk comes from mountain pastures as opposed to valley pastures. Even the location within the same pasture affects the properties of Abondance: the plant species on the north side of one grazing area induced more protein breakdown in the resulting cheeses, which were less firm, more easily fractured, and stickier than cheeses from the south side. The north-side cheeses were also more bitter and salty with more intense aromas.

Soil has an impact on the plants growing in it, and some soil components carry over to the milk. As mentioned in Chapter 10, the salty flavor of Cheshire originates in the soil. Cantal and Salers come from the Cantal Mountains, which are extinct volcanoes, and the cheese is described as having a detectable metallic flavor as a result.

Microorganisms that find their way into the milk and cheese vary with location, as we saw in Chapter 9.

As with many aspects of cheese, disagreement exists over the impact of *terroir*. Professor Paul McSweeney of University College Cork, Ireland, is skeptical about the concept, saying that the rumen of the cow has a great leveling effect. The biological processes occurring in the bovine stomach can produce compounds that override any of the characteristics listed above. Consumers who are not experts in cheese may be unable to detect any differences due to geography. Much depends on your experience and ability to detect subtle flavors.

Now that we have explained the concept of *terroir*, let's move to a chapter and topic that have nothing to do with it: process cheese.

Presidential debates are to real debates as
processed cheese is to cheese.
—GEORGE F. WILL, political columnist, 2004

14

PROCESS CHEESES AND NUTRITION

FOR MANY ADULTS, the problem with process cheeses is that they lack the authenticity, flavor, and texture of the real thing. Although process cheese is frowned upon by connoisseurs, it does provide nutrition, including proteins, vitamins, and minerals. We will discuss process cheese ("processed" is not the legal term in the United States) and nutritional aspects of cheese in this chapter.

Process Cheese

Starting in 1911, James L. Kraft began experimenting with blending and cooking young cheese with aged cheese to obtain a product with a long shelf-life. Kraft had been selling cheese to Chicago grocers since 1903, and was looking for a way to utilize cheese that had not sold. He received a patent for the "Process of Sterilizing Cheese and an Improved Product Produced by Such Process" in 1916, and process cheese was born. In the past, factories used trimmings and older cheese for this procedure, but nowadays much of the cheese is younger stirred-curd Cheddar prepared especially for processing. The average age of the blended cheese is typically three months. The manufacturer may add ingredients from these nine categories:

- Acids such as citric and propionic acid, for preventing microbial growth and improving texture
- Butter oil or cream, for melting

- Color such as annatto and β-carotene
- Emulsifying salts such as sodium phosphate, which promote emulsification of fat and hydration of protein
- Flavoring, including spices
- Preservatives for preventing mold growth
- Skim milk powder and whey powder, to boost the protein content
- Stabilizers (see Box 4.2)
- Sweeteners such as sugar and corn syrup

The product is prepared in a cooker, with grinding, blending, heating, and often homogenization. It must be heated to at least 150° for at least 30 seconds. As the mixture cools, a new matrix forms as caseins interact with each other and enclose emulsified fat globules. The strands of casein are shorter and finer than those of natural cheese, and the fat globules are smaller and more uniformly distributed. The globule size decreases as the processing temperature and concentration of salts increase, which causes the product to become firmer and less meltable.

The United States has legal definitions for process cheese labeling (see Chapter 16 for more on laws regarding cheese). In brief:

- "Pasteurized process cheese" is made from one or more cheese varieties. If the varieties are Cheddar and Colby, it is labeled "Pasteurized process American cheese." If it contains Cheddar, Colby, and other varieties, it is labeled "American cheese." The moisture content is usually no higher than 43 percent, and the fat content is usually at least 47 percent. The pH must be at least 5.3. It cannot contain skim milk or whey powders, stabilizers, or sweeteners.
- "Pasteurized process cheese food" must contain at least 51 percent cheese. The moisture content must be no higher than 44 percent, and the fat content must be at least 23 percent. The pH must be at least 5.0. It cannot contain stabilizers or sweeteners.
- "Pasteurized process cheese spread" must contain at least 51 percent cheese and must be spreadable at 70°. The moisture content must be between 44 percent and 60 percent, with at least 20 percent fat. No pH requirement is specified, and the product may include ingredients from all nine categories listed above.

None of the above can contain cream or Neufchatel cheese. "Pasteurized blended cheese," however, does contain them but cannot include emulsifying salts or acids. Other products on the market are labeled "pasteurized process cheese food," "pasteurized process cheese product," and so on. These are names devised by the manufacturers and are not spelled out in any regulations. Velveeta, for instance, is sold as "pasteurized prepared cheese product." It was developed in 1918 by Emil Frey (the Liederkrantz inventor whom we first met in Box 2.7) and derives its name from "velvet," alluding to its smooth texture. Kraft Foods, which had bought the Velveeta Cheese Company in 1927, used to label the product as "pasteurized process cheese spread." They changed it after the FDA warned them in 2002 that one of the ingredients was milk protein concentrate, which is not included in any of the nine categories listed above.

Cold-pack cheese, also called "club cheese," is a blend of at least one variety each of fresh and aged cheese. It differs from process cheese in that it is manufactured without heating.

Children, who tend to prefer bland foods, are often fed process cheese, and sometimes expect cheese to have the same characteristics when they grow up. In fact, commercially made non-process cheeses in America tend to have fewer and less intense flavors than those from small outfits because large-scale operations can maximize profits by replacing fat and protein with water up to the legal limit. Moreover, higher moisture content could lead to formation of off-flavors from more extensive enzymatic activity, so these cheeses are stored for shorter times, which also results in decreased production of desirable flavors.

Other Products

Some cheeses have added flavors or spices, which are legal as long as the additives do not simulate the flavor of cheese. Fruits, vegetables, and meats are also legal additives. Cheese may also be smoked while being cured at room temperature or with heat, or by adding "liquid smoke" to it. The latter is obtained by burning wood or sawdust, allowing the smoke to collect and condense in a cold vessel, and dissolving it in water. The name of the cheese has to include the name of whatever has been added.

Imitation cheeses and cheese analogues cannot be considered "cheese" since they are manufactured with vegetable oil substituting for milk fat. These products exist because they are less expensive than authentic cheese, are usually not noticed when used as an ingredient (such as in fast food restaurants or ready-to-eat foods), and can be promoted as low in saturated fat. The United States does not have standards for these products, either. Pasteurized process sandwich slices, for example, contain casein and whey along with starch and soybean oil, but no actual cheese.

Nutrition

Cheese, including the process version, is highly nutritious, as it contains protein, vitamins, and minerals that people need. The surviving lactic acid bacteria are considered probiotic, conferring health benefits. Even the fat has some advantageous properties. Consumers purchase products such as cheese in part for their nutritional benefits, even if they do not consciously realize it (Box 14.1).

Table 14.1 shows the major components of cheese along with recommended dietary allowances (RDA) for men and women aged 19–70. RDA is defined as the average dietary-intake level sufficient to meet the nutrient requirements of 97–98 percent of the healthy individuals in a group. The National Academy of Sciences reviews the scientific evidence and periodically sets the RDA for foods. The Academy has not established an RDA for water or for polyunsaturated fatty acids (PUFA), so the numbers listed in those columns are for "adequate intake." No minimums or maximums are stated for fat, cholesterol, or energy, so those spaces are left blank. The information for cheese was compiled by the Food Standards Agency in the United Kingdom and by the U.S. Department of Agriculture (USDA) Nutrient Data Laboratory, and is based on averages for typical cheeses. Within Table 14.1, the quantities listed for RDA are per day, and the other quantities are per 100 grams (3½ ounces) of cheese. The abbreviation for milligrams is mg, and T stands for trace amount. The

BOX 14.1
WHY PEOPLE BUY CHEESE

Rule developing experimentation (RDE) was created by Howard Moskowitz, the President of Moskowitz Jacobs, Incorporated, a market research firm he founded in 1981 in White Plains, New York. In RDE, researchers create short advertisement-like phrases ("vignettes") about a product, present them to consumers, and record their reactions to determine what exactly is appealing. Moskowitz's group provided 36 vignettes about cheese to 241 consumers and found that overall and among males, females, and people under 40, "the classic, traditional flavor of your favorite mozzarella, cheddar, or American cheese" and "the robust and zesty flavor of your favorite aged cheese" were the only high-impact vignettes. Brand names and health messages bore modest or no relevance. But when analyzing the pattern of reactions to cheese ("mindset segmentation") the researchers determined that 58 of the consumers were highly interested in health specifics, 46 more found healthy eating appealing, and the other 137 were most concerned about brand and tradition. The vignettes "An essential source of the nutrients that are important for heart health . . . like potassium, magnesium, and folic acid" and "Contains 13 vitamins and minerals your body needs" carried medium impact among all respondents but had the two highest impacts among those with the health-specifics mindset. "Low fat . . . only 2 g per serving" scored 32nd among all respondents but sixth among the health-specifics consumers. Moskowitz credited the differences in response to allowing people's subconscious feelings to take over. This information gives some direction to manufacturers who wish to appeal to people with a particular mindset.

TABLE 14.1

Cheese components and RDA

	Water (grams)	Protein (grams)	Fat (grams)	PUFA (grams)	Carbohydrate (grams)	Cholesterol (mg)	Energy (Calories)
RDA							
Males	3700	56		18	130		
Females	2700	46		13	130		
Fresh							
Cottage	78.6	12.6	4.3	0.13	3.1	16	101
Cream	45.5	3.1	47.4	1.44	T	95	439
Whey							
Ricotta	72.1	9.4	11.0	0.39	2.0	50	144
Pickled							
Feta	56.5	15.6	20.2	0.59	1.5	70	250
Stretched curd							
Mozzarella	46.5	26.0	20.0	0.51	0.6	54	302
Surface mold							
Brie	48.7	20.3	29.1	0.91	T	93	343
Camembert	54.4	21.5	22.7	0.72	T	72	290
Smear-ripened							
Limburger	48.4	20.1	27.3	0.50	0.1	90	327
Interior mold							

(*continued*)

Table 14.1 (*continued*)

	Water (grams)	Protein (grams)	Fat (grams)	PUFA (grams)	Carbohydrate (grams)	Cholesterol (mg)	Energy (Calories)
Roquefort	41.3	19.7	32.9	1.32	T	90	375
Stilton	45.8	19.9	31.3	1.11	0.1	105	362
Cheddared							
Cheddar	36.6	25.4	34.9	0.98	0.1	97	416
Cheshire	40.6	24.0	31.4	0.87	0.1	90	379
Stirred curd							
Edam	43.8	26.7	26.0	0.67	T	71	341
Gouda	40.4	25.3	30.6	0.86	T	85	377
With eyes							
Emmental	35.7	28.7	29.7	0.97	T	90	382
Gruyère	35.0	27.2	33.3	1.73	T	100	409
Very hard							
Parmesan	27.6	36.2	29.7	1.01	0.9	93	415
Process							
American	39.6	18.1	31.8	1.29	3.7	100	371

Mozzarella is low-moisture part-skim, and the American cheese is pasteurized process without added vitamin D.

The energy is calculated by multiplying the protein and carbohydrates by four (they average 4 Calories per gram), multiplying the fat by nine, and adding the two numbers. The composition of most of the varieties is in this range: 35–50 percent water, 19–29 percent protein, 20–35 percent fat, 0–4 percent carbohydrate, 70–105 mg cholesterol, and 290–450 Calories. You'll note that as the water content increases, the amounts of the other components decrease. The outliers contain low moisture (Parmesan), high moisture (cottage, Ricotta, Feta), or high fat (cream). Now take a brief look at what the components do in humans.

Proteins and Peptides

You can see from the table that cheese is a significant source of protein. About 98 percent of the protein in cheese (except for ricotta and other whey cheeses) is casein. Casein contains every essential amino acid, though the amounts of sulfur-containing cysteine and methionine are relatively low. Much of the protein is broken down to amino acids and peptides before consumption, making cheese highly digestible. Some of the peptides are bioactive, meaning that they exert an effect on the body. The main bioactivity observed is inhibition of angiotensin-converting enzyme (ACE). ACE converts angiotensin I to angiotensin II, a hormone that constricts blood vessels and leads to hypertension. People take ACE inhibitors to lower their blood pressure, but studies on Gorgonzola, Gouda, Parmesan, and others have found ACE inhibitors among the peptides.

Fat and Cholesterol

The vast majority of cheese varieties contain at least 20 percent fat. That is one reason why cheese tastes so good: most of its volatile compounds are soluble in fat and not water, making fat a carrier of flavor. Fat also greatly affects the texture of cheese. About two-thirds of the fat in cheese is saturated, which physicians want their patients to limit in their diets. However, studies including over 347,000 subjects have shown that saturated fat intake is not associated with an increased risk of cardiovascular disease, coronary heart disease, or stroke. Cheese consumption does not increase total and low-density-lipoprotein (LDL) cholesterol concentrations when compared with an equivalent intake of fat from butter. The high calcium content of cheese evidently results in a higher excretion of fat.

You'll remember from Box 5.3 that cheese contains saturated, monounsaturated, and polyunsaturated fat, along with some *trans* fat. Polyunsaturated fat, which is in the form of linoleic and linolenic acid in cheese, is credited with lowering heart-attack risk. In general, the higher the total fat level, the higher the PUFA. Trans fat has been linked to increases in LDL and decreases in high-density lipoprotein (HDL), but research has shown that it is the artificially occurring elaidic acid created in partially hydrogenated oils that is harmful; the naturally occurring vaccenic acid found in dairy products offers a protective effect against coronary heart disease in mice.

As mentioned in Box 5.3, conjugated linoleic acid (CLA) may have anti-carcinogenic properties, and 0.3–0.7 percent of the fat in cheese is CLA. CLA is generated from linoleic acid, and linoleic acid by bacteria in the rumen. The amount increases when animals forage on lush green pastures, decreases with lengthy aging due to lipolysis, and may be affected by processing methods during cheesemaking. The CLA content of process cheese is 0.5–1.0 percent of the fat, making it a better source of CLA than natural cheese. When consuming 4 ounces of Cheddar per day for four weeks, the CLA concentration in the blood of nine subjects increased by 19–27 percent. Studies on anticancer properties of CLA in humans have not been completed, but results on mice and rats are promising.

With fat comes cholesterol. Cholesterol is required for building and maintaining cell membranes and is converted to bile in the liver, allowing for absorption of fats and fat-soluble vitamins. Cholesterol is manufactured by the body and obtained through the diet, but high levels in the blood are associated with heart attack and narrowing of the arteries.

Carbohydrates

Lactose-intolerant people can eat ripened cheese because nearly all of the lactose in milk is lost in the whey during cheesemaking, and most of the rest is metabolized by lactic acid bacteria. Only the higher-moisture cheese varieties, which hold some whey, contain more than a tenth of a percent of lactose.

Vitamins

Next are the vitamins (Table 14.2), which fall into two groups. The fat-soluble vitamins, those that are stored in fatty tissues and the liver, are A, D, E, and K. The water-soluble vitamins, which are not stored in the body and must be replenished daily, are C, which is almost nonexistent in cheese, and the B-complex vitamins. The latter

TABLE 14.2

Vitamins in cheese

Vitamins:	A	D	E	K	Thia	Ribo	Nia	Pan	B_6	Bio	Fol	B_{12}
RDA												
Males	900	15	15000	120	1200	1300	16000	5000	1300	30	400	2.4
Females	900	15	15000	90	1100	1100	14000	5000	1300	30	400	2.4
Fresh												
Cottage	46	0	100	0	50	240	200	300	50	5.1	22	0.6
Cream	385	0.3	1000	2.9	30	130	100	270	40	1.6	11	0.3
Whey												
Ricotta	185	0.2	30	1.1	20	190	100	NR	30	NR	12	0.3
Pickled												
Feta	220	0.5	370	1.8	40	210	190	360	70	2.4	23	1.1
Stretched curd												
Mozzarella	160	0.4	330	1.3	101	329	120	NR	79	NR	10	2.3
Surface mold												
Brie	297	0.2	840	2.3	30	330	500	500	140	3.6	55	0.6
Camembert	230	0.1	650	2.0	50	520	900	800	230	7.5	83	1.1
Smear-ripened												
Limburger	340	0.5	230	2.3	80	503	158	NR	86	NR	58	1.0

(*continued*)

Table 14.2 (*continued*)

Vitamins:	A	D	E	K	Thia	Ribo	Nia	Pan	B_6	Bio	Fol	B_{12}
Interior mold												
Roquefort	295	NR	550	NR	40	650	570	500	90	2.3	45	0.4
Stilton	360	0.2	600	NR	30	470	700	900	130	3.3	78	1.2
Cheddared												
Cheddar	364	0.3	520	2.8	30	39	10	500	150	4.4	31	2.4
Cheshire	350	0.2	700	NR	30	480	100	310	90	4.0	40	0.9
Stirred curd												
Edam	188	0.2	480	2.3	30	350	100	380	90	1.8	40	2.1
Gouda	258	0.2	570	2.3	30	300	100	320	80	1.4	43	1.7
With eyes												
Emmental	320	0.5	440	2.5	50	350	100	400	90	3.0	20	2.0
Gruyère	325	0.3	580	2.7	30	390	40	350	110	1.5	12	1.6
Very hard												
Parmesan	371	0.3	760	1.7	30	320	100	430	110	3.3	12	3.3
Process												
American	254	0.6	270	3.4	30	350	70	NR	70	NR	8	0.7

group includes thiamin (Thia), riboflavin (Ribo), niacin (Nia), pantothenic acid (Pan), vitamin B_6, biotin (Bio), folate (Fol), and vitamin B_{12}. The numbers for pantothenic acid, biotin, and K are for "adequate intake." The amounts are micrograms per day for the RDA numbers and micrograms per 100 grams of cheese for the other data. The vitamin K figures are from the Nutrient Data Laboratory. NR stands for "not reported."

Vitamin A (retinol) promotes vision and keeps mucous membranes and skin healthy. β-carotene, described in Box 2.5, is converted to vitamin A in the intestine and liver. Eating 3½ ounces of most of these varieties will give you at least a quarter of your RDA of vitamin A. Vitamin D (calciferol) regulates blood levels of calcium and phosphate, thereby promoting bone and tooth health. Vitamin E (tocopherol) is an antioxidant that helps prevent cell injury. Vitamin K (phylloquinone) is necessary for blood clotting.

Thiamine (also called vitamin B_1), helps with energy metabolism and nerve function. Riboflavin (also known as vitamin B_2), and pantothenic acid are involved in the metabolism of fats, carbohydrates, and proteins. Cheese is considered a good source of riboflavin. Niacin (a.k.a. vitamin B_3), functions with some cellular enzymes and has a role in cholesterol reduction. Vitamin B_6, pyridoxine, is partially responsible for amino acid metabolism. Biotin helps produce and metabolize fats. Folate is used to make and repair DNA. Vitamin B_{12}, cobalamin, is produced by bacteria, including those in cheese. B_{12} helps to create new cells and maintain nerve cells, and cheese provides a good deal of it.

Minerals

And here in Table 14.3 are the minerals, in units of milligrams per 100 grams of cheese, along with RDA in units of milligrams per day for men and women aged 19–70. The abbreviations for the chemical elements are shown here. The potassium (K) and sodium (Na) data are for adequate intake. For women aged 51–70, the RDA for calcium (Ca) increases to 1200 mg per day, and the RDA for iron (Fe) decreases to 8 mg per day. The other minerals shown here are copper (Cu), iodine (I), magnesium (Mg), phosphorus (P), selenium (Se), and zinc (Zn). As before, NR stands for "not reported" and T is "trace amount."

Cheese is considered a good source of calcium, magnesium, and phosphorus. Calcium is involved in blood clotting, magnesium and phosphorus are necessary for enzyme reactions and transporting cellular energy, and all three are required for forming bones and keeping them strong and healthy. These minerals, along with potassium and sodium, play roles in water-balance and blood vessel, muscle, and nerve activity. Zinc is part of some proteins and has a role in enzyme activity. Iron is used for oxygen transport (especially in hemoglobin, found in red blood cells), copper is an antioxidant,

TABLE 14.3

Minerals in cheese

Minerals:	Ca	Cu	I	Fe	Mg	P	K	Se	Na	Zn
RDA										
Males	1000	0.9	0.15	8	420	700	4700	0.055	1500	11
Females	1000	0.9	0.15	18	320	700	4700	0.055	1500	8
Fresh										
Cottage	127	T	0.024	T	13	171	161	0.004	300	0.6
Cream	985	0.04	NR	0.10	10	100	160	0.004	300	0.5
Whey										
Ricotta	240	T	NR	0.40	13	170	110	NR	100	1.3
Pickled										
Feta	360	0.07	N	0.20	20	280	95	0.005	1440	0.9
Stretched curd										
Mozzarella	731	NR	NR	0.25	26	524	95	NR	652	3.1
Surface mold										
Brie	256	T	0.016	T	15	232	91	0.005	556	2.0
Camembert	235	T	NR	T	14	241	104	0.007	605	2.1
Smear-ripened										

Limburger	497	NR	NR	0.13	21	393	128	NR	800	2.1
Interior mold										
Roquefort	530	0.09	NR	0.40	33	400	91	NR	1670	1.6
Stilton	326	0.04	0.040	0.20	15	314	96	0.007	788	2.9
Cheddared										
Cheddar	739	0.03	0.030	0.30	29	505	75	0.006	723	4.1
Cheshire	560	0.13	0.046	0.30	19	400	87	0.011	550	3.3
Stirred curd										
Edam	795	T	0.013	0.30	34	508	89	0.007	996	3.8
Gouda	773	T	NR	0.30	32	498	85	0.008	925	3.9
With eyes										
Emmental	950	0.13	NR	0.30	18	400	91	0.007	450	1.6
Gruyère	970	0.13	NR	0.30	37	610	99	NR	670	2.3
Very hard										
Parmesan	1025	0.84	0.072	0.80	41	680	152	0.012	756	5.1
Process										
American	1045	NR	NR	0.63	26	641	132	NR	1671	2.5

and iodine and selenium are found in thyroid hormones, but cheese has only minor amounts of those minerals. A study has shown that replacement of dairy products with other foods containing calcium adversely affects the intake of magnesium and phosphorus, as well as protein, potassium, and vitamins A, B_2, B_{12}, and D. Though you can consume enough calcium without eating dairy foods, calcium-replacement foods are not nutritionally equivalent unless you eat unrealistic amounts.

Other Health Benefits

Some people feel that dairy products are bad for you (Box 14.2). But milk, cheese, yogurt, and so forth are complex foods containing many nutrients, so singling out cholesterol or saturated fats as being dangerous to one's health ignores the possibility that other components may in fact be beneficial. The only way scientists can draw conclusions about health and a particular food is by carrying out long-term studies of the consumption of that food and seeing how disease and death rates are affected. Though consumption of cheese and dairy products has long been implicated as increasing the

BOX 14.2
ON THE OTHER HAND . . .

> *"We recommend deleting milk and dairy products from your grocery list or taking them in extremely small quantities. You can make up the minerals by taking bone meal and the B vitamins with brewer's yeast."*
> —J. I. RODALE, *The Complete Book of Food and Nutrition*, 1961

You wouldn't be reading this book if you ate bone meal and brewer's yeast instead of cheese and other dairy products. Rodale was concerned about antibiotics finding their way into milk, but sick cows are now taken out of the milk supply. Farmers can tell if a cow is ailing by checking the milk for the number of somatic cells, the white blood cells that indicate that a cow is fighting an infection. Analyses are always run on every truckload of milk, and if trace amounts of antibiotics are detected, which happens a few times a year across the United States, that milk is then dumped. Tests for six common antibiotics such as penicillin take just minutes, and the FDA is looking for rapid techniques for other antibiotics (the current tests take a week for results to appear and the milk would be in stores by then). The institute Rodale founded is now working with our laboratory on the properties of organic milk from pasture-fed cows.

Some people just don't like cheese. The conversion of milk by microorganisms into a solid product is accompanied by the liberation of strong-smelling compounds. Some of these compounds are also generated when microbes spoil food, work in the digestive system, consume dead animals, and do other things that microbes do. To avoid food

BOX 14.2 (*continued*)

poisoning, people are taught to shun decay and some are sensitive to the aromas surrounding it, causing them to shy away from safe products created by controlled spoilage.

Others do not eat cheese by choice. Vegetarians do not eat meat, fish, or poultry. Lacto-vegetarians eat dairy products but not eggs, ovo-vegetarians eat eggs but not dairy products, ovo-lacto-vegetarians eat both, and vegans are people who won't eat any product of animal origin. Some vegetarians will not consume cheese made with animal rennet, considering it to be a meat product. Microbial or vegetable rennets are acceptable to them, and some artisanal and industrial cheesemakers market their products for this group of vegetarians.

Finally, some people restrict their cheese consumption because of possible adverse health effects. We explained in this chapter that these worries may be overblown, but in any case it is always best to consume food in moderation. Even if a food is good for you, too much of it isn't. And foods that are not good for you are tolerable in small amounts. Remember that cheese consumption continues to climb—as does life expectancy.

risk of cardiovascular disease, recent reviews of long-term studies have indicated otherwise. Cholesterol levels in blood did increase modestly in subjects who consumed cheese, but the calcium present also reduced blood pressure significantly. A review of research showed no clear evidence that consuming dairy food results in a higher risk of cardiovascular disease. In fact, some evidence indicates that cheese is associated with reduced risk, which may have to do with the calcium and some types of fat in the cheese. Moreover, feeding volunteers a diet rich in dairy foods and calcium has been found to enhance weight loss and loss of fat from the abdomen; calcium supplements were not as effective. In a study aimed at altering the course of age-related cognitive decline, the results indicated that people who consumed dairy products at least once a day scored higher in memory and other tests. The reason for this result is not yet known, but the theories include the effect of bioactive peptides in whey proteins and the benefits of calcium, magnesium, and phosphorus as described above. Cheese and other dairy foods may reduce the incidence of type 2 diabetes through a mechanism that has not been determined. Dental caries, popularly called tooth decay, is significantly reduced by eating cheese. Four theories for this effect are: cheese inhibits the acid-producing bacteria that cause caries, saliva lessens the effect of the acid, increased salivary flow from eating cheese removes carbohydrates that the acid-producing bacteria thrive on, or calcium and phosphate in the cheese restore tooth enamel. Some combination of the four may be true.

Now that we have described the components of cheese, we can turn to Chapter 15 and find out how they are measured.

Never commit yourself to a cheese without having first examined it.

—T. S. ELIOT, 1935

15

ANALYSIS AND FLAVOR COMPARISONS

SCIENTISTS EXAMINE CHEESE to determine its characteristics and to offer ways of improving it. Composition, microbiology, structure, texture, color, melt, and flavor are all analyzed, and not many foods have been researched in such detail. In this chapter we will see how these classes of analysis are performed and who does them.

Composition

The first step in characterizing a cheese is to determine its composition. Some of the procedures are performed using an instrument that emits light in the near-infrared region; the water, fat, protein, and lactose molecules absorb the light at specific wavelengths in proportion to their concentrations in the cheese. The traditional methods also work well, and here are some of them:

- The moisture content is found by drying a sample in an oven and seeing how much weight is lost from water evaporation.
- Fat content can be determined by mixing pieces of the sample with water and sulfuric acid in a special glass bottle with a long, narrow neck. The acid dissolves the protein and the heat generated melts the oil, which floats on top. The bottle is placed in a centrifuge, which spins it around rapidly, and the amount of fat is measured in the neck.

- The NaCl concentration is found when the chloride is reacted with a silver compound, producing a color or a change in the ability to conduct an electric current.
- The pH is determined with an electrode connected to a meter that measures voltage and converts the number to a pH value. Also, carbohydrate, specifically lactose, is sometimes measured by electrode and meter, with the voltage depending on the extent of a reaction with an enzyme.
- High temperatures are used to measure the protein and mineral content. The protein concentration is found by heating a sample to over 1500°, which converts the nitrogen in the amino acids to nitrogen gas. The electrical resistance in a heated filament changes when the nitrogen gas passes through, with the change being proportional to the amount of protein in the sample. The minerals present in a cheese are found by first heating a sample in a furnace at 1000° to drive off the water, fat, lactose, and protein. The remaining ash is dissolved and sent into an instrument inside which the liquid is sprayed into a flame, where each mineral emits light at characteristic wavelengths. The intensities of the responses reveal the concentrations of the minerals.

More specialized methods are employed to find other components. Scientists use electrophoresis (el-lek-troh-for-EE-sis, movement of particles by an electrical force) when looking at specific proteins, such as α_{s1}-casein and β-casein. The proteins are extracted from the cheese with water and broken apart with urea or sodium dodecyl sulfate. Urea is produced in the body and excreted in the urine, and "SDS" (also called sodium lauryl sulfate) is a detergent found in floor cleaner, shampoo, and toothpaste. A small amount of the extract is placed near the bottom of a thin, flat gel about 1.5 inches square. The gel is subjected to an electrical field, and the different proteins migrate toward the top of the gel at a speed depending on their size and electric charge. The gel is removed, a dye is added, and short stripes (known as bands) appear. The positions of the bands tell the researcher which protein is which, and the amount of dye on the stripe indicates the concentration of that protein. *Peptides*, which are sections of proteins, are examined at the same time, revealing the extent of proteolysis that has occurred.

The fatty acids are analyzed by converting them to esters by dissolving the sample in a solvent containing an acid and a compound that has a methyl group (such as methanol or sodium methoxide). The acid causes the fatty acids to break off of the triglycerides, and the methyl group reacts with the fatty acids to form methyl esters. These fatty acid methyl esters (FAMEs) are injected into a gas chromatograph, which vaporizes the sample in an inert gas and sends it through a column, which is actually a coiled glass or metal tube filled with an inert solid containing a layer of a specific liquid or polymer. The smaller FAMEs pass through the column faster than the larger ones, and the amounts are measured when the gas travels through a flame. An untreated piece of cheese contains negligible levels of methyl esters, so the measurement reflects just the fatty acid composition. A gas chromatograph is also used for analyzing flavor compounds (Box 15.1).

BOX 15.1
IT GOES UP YOUR NOSE

Flavor compounds are determined by gas chromatography (GC). The flame mentioned above in connection with FAMEs is a type of *detector*, a sensor that is responsive to the compounds being examined. Another type of detector used with GC is the mass spectrometer (MS), which bombards the compound with electrons coming from a filament (comparable to the one inside an incandescent light bulb). The electrons cause the molecule to fragment in a predictable way, and the instrument's software suggests which compounds may correspond to the fragmentation pattern. The scientist decides the most likely one, does the same for the other patterns, and thus compiles a list of what compounds are in the sample and how much of each is present. Whereas GC is capable of looking at one class of compound at a time, such as fatty acids, GC-MS can examine

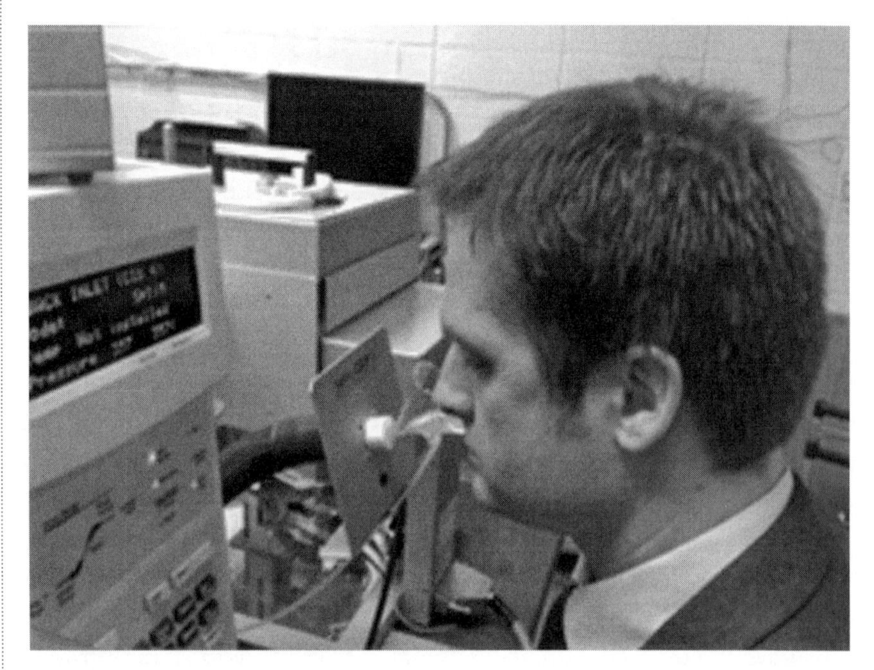

Prof. Devin Peterson of the University of Minnesota detecting cheese aromas using gas chromatography-olfactometry.
Source: Courtesy of Devin Peterson.

BOX 15.1 (*continued*)

various classes. The technique is used in forensics and airport security, and NASA has launched several interplanetary probes with a GC-MS instrument inside.

Another GC detector is the human nose. In this technique, gas chromatography-olfactometry (GC-O), the gas that exits the GC is split, with half going into an MS unit and the other half traveling through a heated tube that is smelled by a scientist (in the photo below, it's Dr. Devin Peterson of the University of Minnesota). As each odiferous compound passes through, the scientist notes the aroma. One thus determines not only which odor-active compounds are present in a sample, but what they smell like in a cheese matrix. GC-O is valuable because it provide guidance to understanding what volatile compounds potentially contribute to the aroma of foods.

Microbiology

A bacteria, fungus, mold, or yeast in cheese is identified by first growing and isolating it. Microorganisms are frequently grown in a petri dish containing agar, a gelatinous extract from red algae. Most microbes do not digest agar but can grow on top of it when it is at the proper temperature, forming visible colonies. A microbiologist may then use genetic techniques to identify the microbe, study the observable traits such as structure and physiological properties, or use molecular characterization, where fatty acids, nucleic acids, proteins, and enzymes are identified. The number of living microorganisms is determined by counting the colonies on a plate, leading to a calculation of colony-forming units per gram.

Structure

The internal structure of cheese used to be hidden from view until the wheel or block was cut open. Some cheese plants have X-ray machines to see if the structure has formed properly. Researchers at the University of California–Davis have examined Swiss cheese with an MRI (magnetic resonance imaging) unit to check eye development (they did this at their laboratory and not at a hospital where patients were waiting). These methods allow the macrostructure to be viewed.

To observe the microstructure, features too small to be seen with the naked eye, a thin slice of cheese is placed in a microscope. A regular optical microscope, using ordinary light, can be used to observe structure and bacteria, but scientists more often use an electron microscope, which is capable of magnifying images by a factor of tens of thousands instead of hundreds. Electron microscopes use electrons emitted by an incandescent electrode (often tungsten) and focused on the specimen, producing black and white images. In scanning electron microscopy (SEM), a rectangular area of a gold-coated specimen is scanned with an electron beam, causing some electrons to be reflected off of the specimen and others to be knocked off of the gold. The result is an image that appears three dimensional. The electron micrographs in this book were obtained by SEM. In transmission electron microscopy (TEM), electrons pass through a very thin slice of sample, producing a two-dimensional image. Certain structural components may be stained with metal to highlight them in the TEM image. We use SEM in our laboratory to view how the bacteria, fat globules, and protein matrix change with aging, and we have observed casein submicelles rearrange during storage using TEM.

Texture

As mentioned in Chapter 10, texture is quite important when enjoying cheese. A common instrumental technique for characterizing the texture of cheese (and many other foods) is texture profile analysis, which is performed on a universal testing machine or a texture analyzer—basically a cheese squeezer. A cylinder of cheese perhaps half an inch in diameter and half an inch high is placed on a flat plate, and a second plate comes down, compresses the sample, comes up, and repeats the action. This test imitates the act of chewing a piece of cheese twice and produces a graph of force *vs.* time that resembles a tall, thin mountain peak followed by a shorter peak. The height of the first peak—the amount of force required to compress the sample in the first "bite"—is the hardness, and the height the sample recovers before the second bite is the springiness. Cohesiveness is the ratio of the area of the second peak to that of the first. If the cheese is sticky, the force needed to pull the upper surface off of the sample is the adhesiveness. If the cheese is brittle, the force at which it breaks is the fracturability. Chewiness is the product of hardness, springiness, and cohesiveness. Rheology (Box 10.4) is characterized by measuring the response when a sample is wiggled gently or twisted until it breaks. These tests provide information on the strength of the protein matrix.

Color and Melting

We saw in Box 2.5 that color is an important factor in acceptability of food. The color of a block of cheese is accurately determined by using a colorimeter. Light is reflected off of the sample, and the relative lightness, redness/greenness, and blueness/yellowness are found. The technique is known as tristimulus reflectance and comes with specific well-defined rules.

The extent to which a piece of cheese melts is found by heating it at a particular temperature for a few minutes and noting how far it spreads out. The amount of free oil, the fat that melts and flows to the surface, can be measured by using a variation of the fat-determination procedure above but without adding the sulfuric acid.

Flavor

As described in Box 15.1, aromas can be analyzed by gas chromatography-olfactometry (GC-O). But if you don't want to rely on an instrument to tell you about the texture or flavor of a piece of cheese, you can assemble a group of people to do it. Texture panels and taste panels consisting of ordinary consumers or trained panelists are assembled for this purpose. If panelists are going to be trained, they have to practice on different samples and be given references to judge them against (a training session is shown in Figure 15.1). For instance, standardized descriptive language for Cheddar flavor—a lexicon—is shown in Box 15.2. Complete training takes perhaps twenty hours, spread over days or weeks. Taste panelists sniff, bite into, and chew the cheeses, inhaling through the nose and from inside the mouth to obtain the orthonasal and retronasal aroma perceptions mentioned in Chapter 7. Texture panels are able to tell quite a bit from manipulating the sample in their fingers before biting into it. Impressions of each component are marked on a scale,

FIGURE 15.1 Prof. MaryAnne Drake of North Carolina State University conducting a cheese taste panel training.
Source: Courtesy of Diane Van Hekken.

BOX 15.2
LEXICON

"I Want Someone to Eat Cheese With."

—TITLE OF MOVIE WRITTEN BY AND STARRING JEFF GARLIN, 2006

The following table (B15.2) contains descriptive language for panelists when they taste Cheddar cheese. The reference materials and compounds in the last column are used to train the panelists about the terms in the first column.

The above was developed just for Cheddar by researchers at Mississippi State University and Sensory Spectrum, Incorporated. Lexicons have also been created for low-fat, full-fat, and smoked Cheddar, Gouda, and Swiss; other aged cheeses; and sheep's milk cheese. The additional terms include such descriptors as ammonia, animal, astringent, biting, buttery, caramel, creamy, decaying animal, fresh fish, goaty, mushroom, pineapple, pungent, rancid, sauerkraut, smoky, soapy, soy sauce, sweaty, and toasty.

TABLE B15.2

Lexicon for Cheddar flavors

	FUNDAMENTAL TASTES	
Term	**Elicited by:**	**Reference**
Bitter	Caffeine, quinine	0.08% caffeine in water
Salty	Salts	0.5% table salt in water
Sour	Acids	0.08% citric acid in water
Sweet	Sugars	5% sucrose in water
	CHEMICAL FEELING FACTORS	
Term	**Definition**	**Reference**
Umami	Elicited by certain peptides and nucleotides	1% MSG in water
Prickle	Sensation of carbonation on tongue	Soda water
	AROMAS AND AROMATICS	
Term	**Associated with:**	**Reference**
Bell pepper	Freshly cut green vegetables	Freshly cut green pepper, 5 ppb methoxypyrazines
Brothy	Boiled meat or vegetable soup stock	Canned potatoes, beef broth cubes, 20 ppm methional

BOX 15.2 (*continued*)

AROMAS AND AROMATICS

Term	Associated with:	Reference
Catty	Tomcat urine	20 ppm 2-mercapto-2-methylpentan-4-one
Cooked	Cooked milk	Skim milk heated to 185° for 30 minutes
Cowy/phenolic	Barns and stock trailers	Bandages, 160 ppm *p*-cresol
Diacetyl	Diacetyl	20 ppm diacetyl
Fecal	Complex protein decomposition	20 ppm indole or skatole
Free fatty acid	Short chain fatty acids	20 ppm butyric acid
Fruity	Different fruits	Fresh pineapple, 20 ppm ethyl hexanoate
Methyl ketone/bleu	Blue-veined cheeses	40 ppm 2-octanone
Milkfat/lactone	Milkfat	Fresh coconut meat, heavy cream, 40 ppm δ-dodecalactone
Moldy/musty	Molds and/or freshly turned soil	Potting soil, 2-ethyl-1-hexanol
Nutty	Different nuts	Lightly toasted unsalted nuts, wheat germ
Oxidized	Oxidized fat	20 ppm 2,4-decadienal
Rosy/floral	Flowers	20 ppm 2-phenethylamine
Scorched	Extreme heat treatment of milk proteins	Milk heated to 250° for 25 minutes
Sulfur	Sulfurous compounds	Boiled mashed egg, struck match
Waxy/crayon	Medium chain fatty acids	10% capric or lauric acid
Whey	Cheddar cheese whey	Fresh Cheddar whey
Yeasty	Fermenting yeast	Raw yeast dough, yeast in warm 3% sucrose water

often 1 to 10, and compared. Water and unsalted crackers are used to cleanse the palate between samples.

Flavor Comparisons

In Chapters 4 through 13 we talked about the classes of cheese and the types of compounds found in them. Now is the time to combine the information to look at the characteristic flavors of cheese varieties.

Researchers used to think that the typical flavor of a cheese variety resulted from just one compound or class of compounds. In the 1950s, Cornell's Frank Kosikowski developed the "component balance theory," which states that cheese flavor was produced by a combination of compounds and that the cheese would not taste right if the compounds were not in the correct proportions. The theory has held up, as scientists have found that many compounds from different categories compose the aroma and flavor of cheese. Some are obviously more important than others, but all contribute.

In the following tables, each column lists a compound and the aroma corresponding to it in that type of cheese. "Farm Ched" is a farmhouse Cheddar, "Grana Pad" is Grana Padano, and "Bov Mozz" and "Buff Mozz" are Mozzarellas made from cow's milk and water buffalo's milk, respectively. The goat cheese is Bouton de Culotte ("trouser button"), a 2-ounce French cheese that becomes hard and brittle with time. "Smear" corresponds to an unspecified French smear-ripened cheese. The odor descriptors were obtained from researchers using GC-O in several different studies.

We start with acetic acid, propionic acid, and the short-and medium-chain fatty acids, which we read about in Chapter 5 (see Table 15.1).

All of the compounds listed for Cheddar and Roncal are essential parts of the odors of those varieties. Butyric acid is a particularly important odorant in Camembert, Grana Padano, Gruyère, Pecorino, and Ragusano, and caproic acid is a key compound in Grana Padano.

And in Table 15.2 are branched-chain fatty acids along with a long-chain fatty acid, lauric.

Isovaleric acid is an important compound in Camembert and Gruyère. 4-Ethyl caprylic acid and 4-methyl butyric acid are the major aroma compounds for the goat cheese listed.

Nitrogen-containing compounds were also described in Chapter 5, and the ones shown here in Table 15.3 are major contributors to the aromas of these cheeses.

You may wonder why anyone would want to eat something that smells like a dirty sock or a stable. Remember that these aromas are combined with many others to form the total picture, and cheese would not taste the same without them.

TABLE 15.1

Short-chain acids in cheese and their aromas

	Acetic	Propionic	Butyric	Caproic	Caprylic	Capric
Cheddar	Peppers, green	Gas, burnt	Toasted cheese	Bad breath, popcorn		Stale butter
Farm Ched	Vinegar		Fecal, cheesy		Sweat	
Emmental	Pungent, vinegar	Pungent, sweet	Rancid			
Goat			Cheesy	Goat	Rancid, pungent	Goat
Grana Pad	Acetic, sour		Cheesy, rancid	Cheese rind, sweaty	Fatty, rancid	Grassy, fatty
Gruyère	Fruity, pungent		Sweaty	Goat		
Parmesan			Cheesy, rotten			
Pecorino			Dirty sock			
Ragusano	Acetic		Dirty sock			
Roncal	Vinegar		Strong cheese, manure	Cheese	Cheese rind	Cheese, milk
Smear	Vinegar, pungent	Vinegar, pungent	Rancid cheese, putrid	Pungent, blue cheese	Goat, waxy	Rancid fat

TABLE 15.2

Lauric and branched-chain fatty acids in cheese and their aromas

	Lauric	Isobutyric	Isovaleric	4-Methyl caprylic	4-Ethyl caprylic
Camembert			Cheese		
Cheddar	Soapy				
Farm Ched			Swiss cheese		
Emmental			Sweet, sweaty		
Goat			Cheese	Goat	Goat
Grana Pad		Cheese, rancid butter	Cheese rind, rancid		
Gruyère		Sweaty			
Roncal	Soap, hot milk		Roncal cheese		

TABLE 15.3

Nitrogen-containing compounds in cheese and their aromas

	2-Ethyl-3,5-dimethyl pyrazine	3-Ethyl-2,6-dimethyl pyrazine	2,3-Diethyl-5-methyl pyrazine	Indole	Skatole
Cheddar					Fecal
Gruyère	Earthy		Earthy	Mothball	
Bov Mozz				Stables, jasmine	
Buff Mozz				Stables, jasmine	
Pecorino		Roasty			
Ragusano		Roasty			

Table 15.4 shows the alcohols that we first saw in Chapter 6.

We see that the compounds in the last three columns have a consistent odor, with some differences in the first three columns. These results show that compounds can be perceived as having various aromas, depending on the other compounds present. 2-Heptanol is an important odorant in Gorgonzola and Grana Padano, and 1-octen-3-ol

TABLE 15.4

Alcohols in cheese and their aromas

	Pentanol	3-Methyl butanol	2-Heptanol	1-Octen-3-ol	Phenyl ethanol	2-Phenyl ethanol
Camembert			Sweet	Mushroom		Floral
Cheddar	Fruity	Plastic				
Farm Ched						Rosy
Goat				Mushroom	Rose	
Gorgonzola	Alcoholic	Pomace	Herbaceous	Mushroom		
Grana Pad	Balsamic		Earthy	Mushroom		
Gruyère						Honey
Bov Mozz		Alcoholic		Mushroom	Rose	
Buff Mozz				Mushroom	Rose	
Pecorino			Fruity			
Roncal						Roses
Smear		Fruity, alcohol				Rose

is a key compound in Camembert, Gorgonzola, Grana Padano, and water buffalo Mozzarella.

Next (Table 15.5) are the sulfur-containing compounds, which were covered in Chapter 7.

The perceived odors show a great deal of consistency. Methional is the primary flavor compound in potato, and the sulfides are major odorants in garlic and onion. Methional is a crucial component of the aromas of all the cheeses on this list. Dimethylsulfide is a key compound for Camembert and Cheddar, dimethyldisulfide is important for Ragusano, and dimethyltrisulfide is a necessary part of Gruyère aroma.

Now for the esters (Table 15.6). We saw in Chapter 8 that esters have fruity aromas, and research has shown that to be true in cheese.

Unlike the other classes of compounds above, esters almost always impart a particular aroma without regard to cheese variety. Ethyl butanoate and ethyl hexanoate are key odorants in Cheddar, Gorgonzola, Grana Padano, Pecorino, and Ragusano, as well as Emmental. Ethyl-3-methyl butanoate is important for bovine Mozzarella.

The ketones (Table 15.7) were in Chapter 9.

2-Heptanone and 2-nonanone are vital compounds in Gorgonzola aroma, 2-nonanone is a primary compound for Ragusano odor, and 2-undecanone is important for Camembert. Diacetyl and 1-octen-3-one, which consistently have mushroom and buttery aromas, are essential for most of the varieties in which they are present.

In Table 15.8 are the aldehydes, which we covered in Chapter 10. The flower in the last column refers to Spanish broom, an ornamental plant.

Nonanal is green, fatty, or soapy, and is a key odorant for Grana Padano and water buffalo Mozzarella. 3-Methylbutanal is important for Camembert, Emmental, and Gruyère. Phenylacetaldehyde, which smells like honey or flowers, is another significant contributor to Gruyère aroma.

The lactones from Chapter 11 are in Table 15.9.

The lactones impart coconut and peach aromas to some cheeses. δ-Decalactone appears to be important for Camembert and Emmental flavor.

Table 15.10 shows the three furans that we covered in Chapter 12.

Furans are responsible for a range of odors. Furaneol and homofuraneol are particularly significant compounds for Cheddar and Emmental flavor.

Finally, the terpenes of Chapter 13 appear in Table 15.11.

Some cheese varieties have more terpenes than these three. For instance, Bitto (Chapter 13) contains α-pinene, limonene, and five other terpenes. The French smear-ripened Mont d'Or, also known as Vacherin du Haut-Doubs, is encircled with a strip of spruce and picks up as many as a dozen terpenes, including α-terpineol, limonene, and linalool, from the wood. Several unidentified terpenes with a dirt or fungus odor have been detected in blue cheeses.

TABLE 15.5

Sulfur-containing compounds in cheese and their aromas

	Methanethiol	Methional	Dimethyl sulfide	Dimethyl disulfide	Dimethyl trisulfide
Camembert	Sulfur	Boiled potato	Sulfur	Sulfur, sour	Vegetable, sulfur
Cheddar	Sulfur	Boiled potato	Cabbage		Putrid
Farm Ched		Baked potato			Sulfur, cabbage
Goat		Potato			
Gorgonzola		Boiled potato			
Grana Pad		Cooked milk			
Gruyère	Sulfur	Boiled potato	Sulfur	Sulfur	Sulfur, cabbage
Parmesan	Sulfur	Potato			Garlic
Pecorino		Potato		Onion	Garlic
Ragusano		Potato		Onion	Garlic
Roncal		Boiled vegetables			
Smear		Cooked cabbage	Cabbage	Cauliflower, garlic	Garlic/onion

TABLE 15.6

Esters in cheese and their aromas

	Ethyl acetate	Ethyl butanoate	Ethyl hexanoate	Ethyl octanoate	Ethyl-3-methyl butanoate	Phenylethyl acetate
Camembert			Fruity, malty			Floral, rosy
Cheddar	Fruity, pineapple	Sweet, fruity	Fruity, melon			
Farm Ched				Fruity		
Goat		Fruity	Moldy			
Gorgonzola		Fruity, green	Fruity, apple			
Grana Pad	Fruity, melon	Fruity, green	Fruity, apple	Green, fruity		
Gruyère				Fatty, fruity		
Bov Mozz		Fruity			Fresh cheese	
Buff Mozz		Green, tallowy	Fruity, green			
Parmesan	Fruity, melon					
Pecorino		Fruity, sweet	Orange	Orange		
Ragusano		Fruity, sweet	Orange			
Smear	Fruity, pineapple					Floral, rose

TABLE 15.7

Ketones in cheese and their aromas

	2-Heptanone	2-Nonanone	2-Undecanone	1-Octen-3-one	Diacetyl	Acetoin
Camembert		Malty, fruity	Floral	Mushroom	Buttery	
Cheddar				Mushroom	Buttery	Sour milk
Farm Ched				Mushroom		
Gorgonzola	Gorgonzola cheese	Gorgonzola cheese	Musty			
Grana Pad	Fruity, fatty					
Gruyère	Green					
Buff Mozz	Animals, blue cheese	Hot milk				Herbaceous
Pecorino				Mushroom		
Ragusano		Hot milk		Mushroom		
Smear	Blue cheese				Buttery	

TABLE 15.8

Aldehydes in cheese and their aromas

	Acetaldehyde	Nonanal	2-Methyl propanal	2-Methyl butanal	3-Methyl butanal	Phenyl acetaldehyde
Camembert	Green		Green	Green	Green	Honey
Cheddar	Sweet	Green, fatty	Floral	Caramel		Rosy
Gorgonzola					Herbaceous	Flower
Grana Pad		Green, tallow				
Gruyère	Green		Malty	Malty	Malty	Honey
Bov Mozz		Green, tallow				Flower
Buff Mozz		Green, tallow				Flower
Pecorino		Soapy	Fruity			
Ragusano		Soapy				Daisy
Smear					Green, malty	

TABLE 15.9

Lactones in cheese and their aromas

	δ-Octalactone	δ-Decalactone	γ-Decalactone	δ-Dodecalactone	γ-Dodecalactone
Camembert	Coconut	Coconut			
Cheddar		Coconut		Sweet, soapy	
Farm Ched		Peach, coconut	Coconut	Cheese, coconut	
Goat	Fruity				
Grana Pad		Creamy, coconut		Coconut, buttery	
Gruyère		Sweet, coconut			
Bov Mozz	Peach				
Buff Mozz	Peach				
Parmesan		Sweet, coconut			
Roncal					Peach, almonds

TABLE 15.10

Furans in cheese and their aromas

	Furaneol	Homofuraneol	Sotolon
Cheddar	Caramel	Caramel	
Farmhouse Cheddar	Sweet, burnt sugar		Curry, seasoning
Grana Padano		Caramel, burnt sugar	
Gruyère	Strawberry		Seasoning
Parmesan		Caramel, burnt sugar	Coriander, curry

TABLE 15.11

Terpenes in cheese and their aromas

	α-Pinene	Geraniol	Linalool
Cheddar	Pine, green		
Farmhouse Cheddar			Sweet, floral
Pecorino			Orange, flower
Ragusano		Flower	

Cheese Scientists

Anyone can compare the flavors and textures of different cheeses and rank them by preference. Some organizations formalize the procedure by sponsoring cheese contests (Box 15.3). When cheese is to be purchased for government programs, USDA graders go to the plant to judge the appearance, color, flavor, and texture of samples, and send them to a lab for compositional and microbial testing (Box 15.4). These traditional tests by human judges are subjective, however, and their opinions vary according to the judges' background, experience, and other factors. The scientific techniques in this chapter are more reliable because they give objective results. The researchers who analyze cheese have backgrounds in animal science, biochemistry, chemistry, engineering, food science, microbiology, and related fields. They develop improved starter cultures, create cheese with reduced fat or sodium, improve packaging and preservation, establish standards, characterize obscure varieties, find better ways to analyze cheese, and verify identities and geographic origins (Box 15.5). They also look at health benefits, figure out how compounds are formed in cheese, identify

BOX 15.3
CHEESE CONTESTS

> *"One cheese differs from another, and the difference is in sweeps, and in landscapes, and in provinces, and in countrysides, and in climates, and in principalities, and in realms, and in the nature of things."*
>
> —HILAIRE BELLOC, *First and Last*, 1911

You can measure the differences in cheese for yourself or rely on experts to make a recommendation. Annual contests provide an opportunity for trained cheese judges to evaluate the products of various companies, score them, and determine winners. The Wisconsin Cheese Makers Association has some 2,300 entries grouped into 40 classes at their World Championship Cheese Contest, which is held in even-numbered years. Cheeses are broken out into different categories. The Cheddar class, for instance, includes mild (aged 0–3 months), medium (3–6 months), sharp (6 months–1 year); aged 1–2 years, aged 2 years or more; bandaged (ripened in a cloth wrapping) mild to medium, and bandaged sharp to aged. In 2010 they had 35 judges from 16 countries, including Nana Y. Farkye, a professor in the Dairy Science Department at California Polytechnic State University in San Luis Obispo (who also supplied much of the information below). Nana and the other judges employ a 100-point scoring system including flavor (45 points), body and texture (30), make up and appearance (15), color (5), and rind development (5). Working on the assumption that each cheese/entry is the best a company has to offer, the judges start with an initial score of 100 from which deductions are made for defects to give the final score. A perfect score of 100 is an exception, while scores less than 95 are unusual. Two judges independently score each cheese, and the final score for each entry is the average score of the two judges. The deductions for defects are on a one-tenth point basis: Very slight defect—detected under very critical examination, 0.10 to 0.50; Slight defect—detected upon critical examination, 0.50 to 1.50; Definite defect—detected easily, but not intense, 1.60 to 2.50; and Pronounced—detected easily and intense, 2.60 or greater. Because each judge is experienced, the pair working together may confer with each other initially at the beginning of judging each class to make sure they are seeing the same defects in each entry.

The American Cheese Society has around 100 categories and over 1,000 entries at their annual meeting. A technical judge (who looks at body, flavor, and texture) deducts points from 50, and an aesthetic judge (appearance, aroma, flavor, texture) adds points up to 50. A suggested score card for Cheddar by the American Dairy Science Association lists these flavor defects: bitter, feed (tastes of animal feed), fermented (vinegary), flat/low flavor, fruity, heated (reminiscent of spoiled milk), high acid, oxidized, rancid, sulfide, unclean (dirty aftertaste), whey taint, and yeasty. Their body and texture defects are corky (dry and hard), crumbly (falls apart), curdy (rubbery), gassy (eyes), mealy (grainy), open (spaces in the interior), pasty (sticky), short (flaky), and weak (soft). Appearance defects are not listed but may include mold, mottled, off-colors, and white specks. The specks would be the crystals and NaCl deposits mentioned in the "sound" section of Chapter 6.

BOX 15.4
GRADING CHEESE

USDA graders come into cheese plants or warehouses to give their unbiased opinions on the products. They examine the condition of the cheese container and compare the weight marked on it with the actual weight. Then they use a trier like the one shown in Chapter 3 to extract a cylindrical "plug" of cheese and examine it by sniffing, touching, and tasting. They also cut the cheese in half and examine the two cut surfaces. Any defects are noted, along with the descriptors "very slight," "slight," and "definite." The type and extent of defects are compared with the standards for that cheese variety, and a grade is then assigned. For Swiss-type cheese, the grades are A, B, and C; and for Cheddar, AA, A, B, and C. Barrels containing at least 100 pounds of American cheese are assigned U.S. Extra Grade, U.S. Standard Grade, or U.S. Commercial Grade. Any cheese scoring below the requirements would be listed as "No final U.S. grade assigned." To quote from one standard:

> The determination of U.S. grades of Swiss cheese shall be on the basis of rating the following quality factors: (1) Flavor, (2) Body, (3) Eyes and texture, (4) Finish and appearance, and (5) Color . . . in a randomly selected sample representing a vat of cheese . . . The final U.S. grade shall be established on the basis of the lowest rating of any one of the quality factors.

The eye and texture defects for Swiss-type cheeses were described in Box 12.2, and we'll use that information to look at the standard for Grade B Swiss and Emmental cheese:

> The cheese shall possess well-developed round or slightly oval-shaped eyes. The majority of the eyes shall be 3/8 to 13/16 inch in diameter. The cheese may possess the following eye characteristics to a very slight degree: dead eyes and nesty; and the following to a slight degree: dull, frogmouth, one sided, overset, rough, shell, underset, and uneven. The cheese may possess the following texture characteristics to a slight degree: checks, picks and streuble.

Grade A cheeses are allowed very slight dull, rough, shell, checks, picks and streuble, and none of the other defects. Grade C cheeses would have slight or definite levels of any or all of the defects in Box 12.2.

microbes, improve manufacturing processes, and study structure, texture, and flavor. Serious research on cheese began in the latter part of the nineteenth century after Louis Pasteur, who discovered the handedness in molecules (Chapter 5), also showed that microorganisms caused many diseases and were responsible for spoiling milk. Gradually, scientists learned that microbes were involved in cheese-ripening, enabling

BOX 15.5
THE CHEESE CAPER

> *"We sat through two and a half hours of a fantastic murder mystery that had more holes in it than a piece of Swiss cheese."*
> —MICKEY SPILLANE, *I, the Jury*, 1947

Although most cheese scientists are not hardboiled detectives like Spillane's Mike Hammer, they are sometimes involved in legal matters. As we will see in Chapter 17, the labeling of a number of cheese varieties is regulated, and science is used to track down possible violators. In the late 1980s, the USDA's Agricultural Marketing Service (AMS), which purchases cheese and other commodities for the National School Lunch Program, came to us with a problem. A plant from which AMS purchased Mozzarella was found by inspectors to contain bags of calcium caseinate powder, which could have been added to the milk to increase the yield of cheese inexpensively. The regulations prohibit the use of milk powder in making genuine Mozzarella; if added, the cheese would have to be labeled as imitation. AMS suspected that the plant was selling imitation cheese to them but had no way of proving it. We found that the composition of their cheese was no different from authentic Mozzarella's, and the pattern of protein breakdown was also the same. But then I looked at the melting characteristics of their product and found something wrong. I used a differential scanning calorimeter (DSC), which measures the heat taken in or given off (the "calorimeter" part) by a sample in a tiny pan as it is heated or cooled ("scanning") and compared to an empty pan ("differential"). When the cheese was removed from the refrigerator, and heated immediately in the DSC, the amount of fat that melted at 60°–64° was too low. We made some Mozzarella containing 1% and 2% calcium caseinate and saw that the amount of fat melting in that temperature range decreased. The caseinate apparently acted as an emulsifier and prevented some of the fat from solidifying in the refrigerator, meaning that it would already be liquid when the cheese was heated, producing a smaller DSC signal. We then looked at microstructure with an SEM and saw that the fat globules in genuine Mozzarella were uniformly dispersed, but were often coalesced into large globules in the company's cheese and in the 1% and 2% calcium caseinate cheeses. With this information, we concluded that the company was making imitation Mozzarella and passing it off as genuine. The company, which had been taken to court, changed their plea from not guilty to guilty. They were convicted, fined $515,000 for defrauding the U.S. Government, and went out of business.

Scientists use other advanced techniques to determine if a cheese is really from the place identified on the label. They will see if the microorganisms match those in the place of origin and do the same with ratios of the chemical elements present. If they observe bacteria that are not found in the starter culture or NSLAB of a variety made in a particular area, then the cheese was probably made elsewhere. Similarly, the geology of a region will lead to proportions of certain elements being present in the feed and eventually the cheese.

BOX 15.5 (*continued*)

A major issue with some varieties is the substitution of goat, sheep, or water buffalo milk by cow milk, which is less expensive. Electrophoresis and chromatography are often used to confirm authenticity. ELISA (enzyme-linked immunosorbent assay) is a test in which enzymes and antibodies or antigens are used in a reaction to produce a color change. Specific proteins found in the milk of various species are identified in this manner. ELISA is most often used to detect virus antibodies, some drugs, and food allergens. In another identification technique, PCR (polymerase chain reaction), a section of DNA is separated into its two strands, an enzyme (DNA polymerase) is used to synthesize two-stranded DNA from each strand, and the process is repeated until many copies of the original are formed. PCR is now a mainstay of forensics and disease diagnosis. ELISA and PCR are not routine procedures, but manufacturers may resort to them if they are losing business to competitors who are adulterating their products.

BOX 15.6

IMPERIAL RUSSIA AND CHEESE SCIENCE

Two noted scientists who lived in the Russian Empire were involved in cheese and cheese cultures.

Dmitri Mendeleev (1834–1907) was a chemist who developed the periodic table of the elements and had element 101, Mendelevium, named after him in 1955. He was also a consultant on agricultural matters, including artisanal cheese production, which he thought could be a model for organizing industry. Mendeleev was a member of Russia's Free Economic Society for the Encouragement of Agriculture and Husbandry, a learned society independent of the government that operated from 1765 to 1919. (Luminaries such as Leo Tolstoy were also members.) Mendeleev was supposed to inspect cheesemaking cooperatives for the Society on March 1–12, 1869, but had to delay the trip by a day because his final form of the periodic table was completed and sent to printer on March 1. A colleague presented the periodic table paper to the Russian Chemical Society later that month.

Ilya (also translated as Élie) Metchnikoff (1845–1916) shared the 1908 Nobel Prize in Physiology or Medicine for his research on the immune system. In his later work he incorrectly theorized that aging is caused by slow poisoning by toxins that pathogens produce in the digestive system, and that resistance is weakened as these bacteria proliferate. He concluded that lactic acid bacteria such as those we first mentioned in Chapter 2 would counteract these effects and prolong life. Metchnikoff singled out the thermophile *Lactobacillus bulgaricus* (now called *Lactobacillus delbrueckii* subspecies *bulgaricus*) as having the ability to establish itself in the intestine (it doesn't) and to prevent pathogens there from multiplying (it can't). Others then found that *Lactobacillus acidophilus*, the yogurt culture we first met in Box 4.1, is capable of colonizing the

BOX 15.6 (*continued*)

gastrointestinal tract and exerting a healthful physiological effect. Japanese scientist Minoru Shirota found that a particular strain of another culture used for cheese, *Lactobacillus casei*, also survives in the gut and assists in good intestinal health. In 1935 Shirota began to produce Yakult, a fermented skim milk drink that is quite popular in his native Japan and other places. This product helped accelerate interest in probiotics, live microorganisms that confer a health benefit when adequate amounts are consumed. Some probiotic cheeses have been developed, but the most popular probiotic product that eventually resulted from Metchnikoff's work is commercial yogurt.

BOX 15.7
REPORTING CHEESE RESEARCH

When academic, government, and some industrial scientists complete an aspect of their research, they try to publish a paper about it in a scientific journal. Papers' elements are usually arranged in this order: an abstract, which summarizes the work; an introduction providing background information on the problem being examined, why they are examining it, and the research that has been done previously; materials and methods, describing how the work was performed, with enough detail so that others can repeat it; results and discussion, where the findings are presented and explained along with tables, graphs, and pictures; a conclusion tying everything together; acknowledgements of people and organizations who helped; and a list of articles that the authors cited. Most papers are submitted and published online, which greatly reduces costs. When a paper is received, the journal's editor-in-chief sends it out to at least two scientists in that field who offer suggestions to improve it (though they may reject it altogether). This peer-review process is meant to insure that the quality of the work is sound. The editor has the final decision on whether the paper is good enough to print. A variety of scientific journals publish papers about cheese science, including:

Journal of Agricultural and Food Chemistry
Journal of Food Science
Journal of the Science of Food and Agriculture
Food Chemistry
Food Research International
LWT—Food Science and Technology (the original title was *Lebensmittel-Wissenschaft und Technologie*)
Advances in Food and Nutrition Research
Critical Reviews in Food Science and Nutrition
Trends in Food Science and Technology

Some journals focus on dairy foods. These include:

(*continued*)

BOX 15.7 (*continued*)

International Dairy Journal
International Journal of Dairy Technology
Journal of Dairy Research
Journal of Dairy Science
Dairy Science and Technology (formerly *Le Lait*)

Scientists may opt to publish in other specialized journals that sometimes have articles about cheese, such as *Journal of Food Engineering, Journal of Food Protection, Journal of Texture Studies*, and *Small Ruminant Research.*

Cheesemakers and cheese scientists also read monthly trade journals, such as *Food Product Design* and *Prepared Foods*. Two weekly trade newspapers are also published, *Cheese Reporter* and *Cheese Market News.*

manufacturers to reduce variations within a variety. Two noted Russian scientists contributed to advances in cheese technology while concentrating on other things (Box 15.6).

The goal of research is the dissemination of knowledge by presenting findings at meetings and publishing in scientific journals. In the United States, much of the work on cheese is presented orally or in poster form at meetings of the American Dairy Science Association, the American Society for Microbiology, the Institute of Food Technologists, and other groups. Cheese scientists publish in many of the journals shown in Box 15.7. Some of their work deals with the laws and regulations about cheese, and we will turn to those next.

How can you govern a country in which there are
246 kinds of cheese?
—CHARLES DEGAULLE, *Les Mots du General*, 1962

16

LAWS, REGULATIONS, AND APPELLATIONS

IT IS A challenge to run a nation with so many approaches to making cheese (France now has some 400 distinct varieties), but at least some rules can be enforced regarding how the cheeses are made and labeled. The United States has "Standards of Identity" for cheese and other foods in its Code of Federal Regulations, the rules published in the Federal Register by the government. These rules specify which ingredients may be used in cheese and sets limits on moisture, fat, and storage time. The regulations for cheese are in Title 21 (Food and Drugs), Part 133 (Cheeses and Related Cheese Products). The Food and Drug Administration (FDA) is responsible for enforcing these rules. The laws regarding process cheese, mentioned in Chapter 14, are also found here. The following are the complete regulations for Edam and Gouda, which are found in Sections 138 and 142.

Federal Regulations Concerning Edam and Gouda Cheeses

§ 133.138 EDAM CHEESE

a. *Description.*
   1. Edam cheese is the food prepared by the procedure set forth in paragraph (a)(3) of this section or by any other procedure which produces a finished cheese having the same physical and chemical properties. The

minimum milkfat content is 40 percent by weight of the solids and the maximum moisture content is 45 percent by weight, as determined by the methods described in § 133.5. If the dairy ingredients used are not pasteurized, the cheese is cured at a temperature of not less than 35°F for at least 60 days.

2. If pasteurized dairy ingredients are used, the phenol equivalent value of 0.25 gram of edam cheese is not more than 3 micrograms, as determined by the method described in § 133.5.
3. One or more of the dairy ingredients specified in paragraph (b)(1) of this section may be warmed and is subjected to the action of a lactic acid-producing bacterial culture. One or more of the clotting enzymes specified in paragraph (b)(2) of this section is added to set the dairy ingredients to a semisolid mass. After coagulation the mass is cut into small cube-shaped pieces with sides approximately three-eighths-inch long. The mass is stirred and heated to about 90°F and so handled by further stirring, heating, dilution with water or salt brine, and salting as to promote and regulate the separation of curd and whey. When the desired curd is obtained, it is transferred to forms permitting drainage of whey. During drainage the curd is pressed and turned. After drainage the curd is removed from the forms and is salted and cured. One or more of the other optional ingredients specified in paragraph (b)(3) of this section may be added during the procedures.

b. *Optional ingredients.* The following safe and suitable ingredients may be used:
    1. *Dairy ingredients.* Milk, nonfat milk, or cream, as defined in Sec. 133.3, used alone or in combination.
    2. *Clotting enzymes.* Rennet and/or other clotting enzymes of animal, plant, or microbial origin.
    3. Other optional ingredients.
        i. Coloring.
        ii. Calcium chloride in an amount not more than 0.02 percent (calculated as anhydrous calcium chloride) of the weight of the dairy ingredients, used as a coagulation aid.
        iii. Enzymes of animal, plant, or microbial origin, used in curing or flavor development.
        iv. Antimycotic agents, the cumulative levels of which shall not exceed current good manufacturing practice, may be added to the surface of the cheese.

c. *Nomenclature*. The name of the food is "edam cheese."
d. *Label declaration*. Each of the ingredients used in the food shall be declared on the label as required by the applicable sections of parts 101 and 130 of this chapter, except that:
   1. Enzymes of animal, plant, or microbial origin may be declared as "enzymes"; and
   2. The dairy ingredients may be declared, in descending order of predominance, by the use of the terms "milkfat and nonfat milk" or "nonfat milk and milkfat," as appropriate.

§ 133.142 Gouda cheese

Gouda cheese conforms to the definition and standard of identity and complies with the requirements for label declaration of ingredients prescribed for edam cheese by § 133.138, except that the minimum milkfat content is 46 percent by weight of the solids, as determined by the methods described in § 133.5 and the maximum moisture content is 45 percent by weight.

The second paragraph in the Edam section mentions "phenol equivalents." Milk contains an enzyme, alkaline phosphatase, that is inactivated by pasteurization. This enzyme is more heat-stable than any pathogens that may be present, so the absence of alkaline phosphatase is an indication (though not a guarantee) that the product is safe. The presence of the enzyme is tested by adding disodium phenyl phosphate; the phenyl portion is broken off and reacts with a color-producing compound to produce a measurable blue tint. If a quarter of a gram of Edam produces a value in excess of 3 micrograms of phenol, the cheese is not safe. It is also possible that a cheese failing the test has been adulterated with unpasteurized milk. Unscrupulous manufacturers may resort to adulteration with other ingredients to save money (Box 16.1).

Paragraph (3) of the Edam cheese regulation quoted here provides an outline of the cheesemaking procedure. Some leeway is given—the type of starter culture, length of heating, and pressing conditions are not specified—but moisture and fat levels are strictly defined so that the consumer does not purchase an inferior product. "Antimycotic" under (b) (3) (iv) is another word for *antifungal*.

Appellation Systems

Other countries also have rules and regulations for their cheeses. In 1919, France adopted a standardized system for protecting from imitation any foods traditionally produced in certain regions, as well as assuring consumers that the food

BOX 16.1
ADULTERATION

The first book detailing tainted food was *A Treatise on Adulterations of Food, and Culinary Poisons* published by Fredrick Accum in 1820. Accum was a German chemist who moved to London and made advances in gaslighting and helped popularize science before turning his attention to impure food. Subtitled "Exhibiting the Fraudulent Sophistications of Bread, Beer, Wine, Spiritous Liquors, Tea, Coffee, Cream, Confectionery, Vinegar, Mustard, Pepper, Cheese, Olive Oil, Pickles, and Other Articles Employed in Domestic Economy," the book's cover featured a spider on a web and the quotation "There is death in the pot" from 2 Kings 4:40. He mentioned in the chapter "Poisonous Cheese" that there were cases in which the orange color of the rind of Gloucester cheese was provided by annatto spiked with red lead. The angry reaction from businesses chased Accum back to Germany, but the public became aware of the unethical and dangerous practices occurring in their food supply.

The anti-adulteration cause in the United States was taken up by Harvey Washington Wiley, who was born in a log farmhouse in Indiana in 1844 and began serving as Chief Chemist at the USDA's Bureau of Chemistry, the forerunner of the U.S. Food and Drug Administration, in 1883. Four years later, a report by the bureau noted that the cream in cheesemilk was sometimes replaced with cottonseed oil and margarine, and that arsenic and zinc sulfate could be added to give mild cheese the flavor of aged cheese. Adulteration with mashed boiled potatoes was not unknown. The report also noted that in some places Limburger and other cheeses were soaked with urine to give the appearance of ripeness. Through Wiley's efforts, the Pure Food and Drug Act of 1906 was signed into law by President Theodore Roosevelt, and the Bureau of Chemistry began to conduct the first federal food and drug inspections the next year.

Adulteration has never really stopped, despite the advances in detection by scientists and government agencies and heavy penalties imposed by the courts. In a legendary 1969 case, a product sold as grated Parmesan actually consisted of grated umbrella handles.

While we are on the subject of breaking the law, a 2011 retail survey showed that cheese is the most-stolen product from stores around the world. It had a global theft rate of 3.1%, ahead of fresh meat and candy (both 2.8%).

was true to type. They called it *Appellation d'Originie Contrôlée*, "controlled designation of origin," and Roquefort became the first AOC cheese six years later. Other European countries later developed their own systems. Finally, in 1992, members of what is now the European Union (EU) established three regimes of protected geographical status: protected designation of origin (PDO), protected geographical indication (PGI), and traditional specialty guaranteed (TSG). PDO foods define the (traditional) manufacturing method and usually use the name of a place or area where the food is prepared and produced, and whose quality or properties are significantly or entirely determined by the natural and human environment in that location. PGI is less strict: the food still uses the name of the location where it is traditionally made, but it only has to have some characteristic that is attributed to that location. TSG foods merely have to have traditional composition or production and are not confined to a specific geographical area. Countries outside the EU cannot export into the EU any foods with the same names as protected ones. American cheesemakers, therefore, cannot ship Neufchâtel into the EU—and for good reason, since the American version is low-fat cream cheese and the PDO version is closer to Camembert. (The United States does export and import a lot of cheese, as you can see in Box 16.2). Switzerland, which is neutral and not an EU member, nevertheless abides by EU regulations and uses an AOC system that is equivalent to PDO and is included here.

Here is a portion of the PDO specifications for Xygalo Siteias, the newest Greek cheese to receive PDO status:

> Xygalo Siteias is a product of milk acidification. It is white, pasty and/or granular in texture and skinless. It tastes fresh, sourish, slightly salty, and has a pleasant characteristic aroma. It has a maximum moisture content of 75 percent and a maximum salt content of 1.5 percent, whilst its fat in dry matter ranges from 33 percent to 46 percent and it has a minimum protein content of 31.5 percent.
>
> Xygalo Siteias is produced from goat's milk or sheep's milk, or a mixture of both if the quantity of goat's milk is insufficient, with the fat content of the sheep's milk adjusted, so that the fat content of the final product remains under 46 percent (in dry matter). The milk mixture is then pasteurised (optional) and cooled to 25°C. Salt (NaCl) is added to a maximum of 2 percent by weight, as are harmless acidic bacterial cultures and small amounts of natural rennet from animals' stomachs (mainly if the milk has been pasteurised). Then the milk is left to ferment naturally in food-grade containers that are kept stationary and covered, but not hermetically sealed, for seven to ten days at a temperature of 15°–20°C. The excess fat and butter are removed from the surface of the curd. Ripening continues in these containers for approximately one month at a temperature of 10°–15°C, with no stirring of the curds for the

BOX 16.2
U.S. CHEESE IMPORTS AND EXPORTS

As mentioned in Box 6.2, the United States exported nearly half a billion pounds of cheese and imported 313 million pounds in 2011. Table B16.2-1 lists the leading destinations for American-made cheese in that year.

Half of these countries are in eastern Asia, which shows that their attitudes toward dairy foods are changing. Traditionally, residents of those places did not consume milk, butter, or cheese to a great extent since it was not part of their culture (though dairy products were introduced to Japan when that country modernized in the latter part of the nineteenth century). With increasing globalization, Chinese, Koreans, and others are being exposed to these products, trying them, and liking them. In fact, part of our laboratory's research on Queso Fresco was performed by a visiting scientist from Harbin, China.

The USDA's Foreign Agricultural Service classifies imports by type and country. The leading countries that exported cheese to the United States in the first half of 2012 are listed in Table B16.2-2:

TABLE B16.2-1

Leading exporters of cheese to the U.S.

Type	*Countries*
Blue mold	Denmark, Germany, France
Cheddar	United Kingdom, New Zealand, Australia, Ireland
Edam and Gouda	Netherlands
Gruyère	Netherlands, Germany, Switzerland
Italian type	Argentina, Italy
Swiss and Emmentaler	Finland, Norway, France, Switzerland
All others	France, Denmark, Ireland, Netherlands, Italy

TABLE B16.2-2

Leading importers of American-made cheese

Country	Millions of pounds
Mexico	106.5
South Korea	77.4
Japan	50.5
Canada	24.6
Saudi Arabia	24.3
Australia & Oceania	23.8
Egypt	16.2
China & Hong Kong	18.9

entire duration of the acidification-ripening process. Finally the product is separated from the whey that is concentrated at the bottom of the containers, placed in food-grade casks and refrigerated at a temperature of under 4°C. If the milk has not been pasteurised, Xygalo Siteias should remain refrigerated for at least two months before it is released for consumption so that checks can be carried out to ensure that the product is free from any undesirable micro-organisms.

These regulations are much more detailed than those in the United States, and go on to specify the plants that the animals may feed upon, packaging, exact geographical area of production, and differences between Xygalo Siteias and similar varieties. Any significant variations from a procedure will result in a different cheese, one that is too moist or dry, too hard or soft, and so forth. As we have seen, the *terroir* affects the flavor because of the food that the animals eat, so the geographical limitations are important.

About 200 cheeses carry a PDO, PDI, TSG, or (for Switzerland) AOC appellation. Some cheeses cannot be granted an appellation since they are common enough to be considered generic, so some cheesemakers have to be creative. Cheddar is made in such great quantities around the world that an appellation would be impossible, but handmade West Country Farmhouse Cheddar is a PDO cheese. Lancashire cheese is made all over England, so manufacturers in that county received a PDO for their Beacon Hill Traditional Lancashire Cheese, named after a hill there.

The following sections show PDO cheeses by country. The list highlights an exceptional aspect of cheese: the countless ways to make it. In addition to the parameters listed in Chapter 2, variations include the number of milkings, season, altitude, region, use of alcohol and wood, and even the gender of the cheesemaker. These aspects (except perhaps the gender) lead to a product that is noticeably different than one with a variation in the process. For example, a cheese that is in contact with wood throughout ripening will boast an aromatic balsam flavor that others will not have. Here is where science and art come together.

Under the Characteristics column, C indicates that the cheese is made from cow's milk, G from goat's milk, and S from sheep's milk. Almost all of these cheeses are made from raw milk. Note that some of them are not exported, and a few are rare even in their own countries. The list will undoubtedly be incomplete by the time you read it since additions are made every few months.

Austria

About two-thirds of Austria is mountainous, including the states of Tyrol and Vorarlberg in the west. All of the PDO cheeses are from this area, are made from cow's milk, and have a hard texture. (See Table 16.1.)

France

The French climate may be Alpine, Mediterranean, or oceanic. Much of the land is flat or rolling hills, allowing for cattle grazing, and most of its cheeses are from cow's milk. The country classifies its cheeses into eight categories: fresh, soft with natural rind, soft with washed rind, pressed, pressed and cooked, goat, blue, and processed. (See Table 16.2.)

TABLE 16.1

Austrian PDO cheeses

Name	Translation	Characteristics
Gaitaler Almkäse	Cheese from pastures of the Gailtal (a valley)	Morning milk mixed with skim milk from previous evening
Tiroler Almkäse (or Tiroler Alpkäse)	Cheese from pastures (or Alps) of Tyrol	Washed in brine, smeared with *Brevibacterium linens*
Tiroler Bergkäse	Cheese from mountains of Tyrol	Brined twice a week, develops gray surface mold
Tiroler Graukäse	Gray cheese from Tyrol	Coagulated with lactic acid, pressed twice, develops gray surface mold
Vorarlberger Alpkäse	Cheese from Vorarlberg Alps	Made from summer milk produced at 3300–5900 feet, blend of evening skim and morning whole milk
Vorarlberger Bergkäse	Cheese from Vorarlberg mountains	Similar to Vorarlberger Alpkäse but non-seasonal version and from one milking

TABLE 16.2

French PDO cheeses

Name	Translation	Characteristics
Abondance	Name of cattle breed and town	Abondance C, hard, made immediately after milking, aged 90 days
Abondance de Savoie	Abondance of Savoy	Abondance C, same as Abondance but aged 12 weeks
Banon	Name of town	G, uncooked, unpressed, wrapped in chestnut leaves and then tied with ribbon
Beaufort	Name of town	See Chapter 12
Bleu d'Auvergne	Blue of Auvergne (region)	C, similar to Roquefort, aged at least 2–4 weeks
Bleu de Gex (or Bleu de Haut-Jura or Bleu de Septmoncel)	Blue of Gex (a town), of high Juras (mountains), or Septmoncel (a town)	Montbéilard C, interior mold grows because air is injected with a syringe, outside mold wiped off before eating
Bleu des Causses	Blue of Causses (plateaus)	C, similar to Roquefort, aged at least 70 days
Bleu du Vercors-Sassenage	Blue of Vecors (mountain mass) and Sassenage (a town)	C, interior mold, uncooked, unpressed, mild
Brie de Meaux	Brie of Meaux (a region)	See Chapter 7
Brie de Melun	Brie of Melun (a town)	C, same as Brie de Meaux but coagulated without rennet and aged up to 10 weeks
Brocciu Corse	Corsican fresh goat or sheep cheese	G or S, made from whey to which whole milk is added
Camembert de Normandie	Camembert of Normandy (a town)	See Chapter 7
Cantal (or Fourme de Cantal or Cantalet)	Cheese of Cantal (mountains)	See Chapter 10
Chabichou du Poitou	Goat of Poitiers (a town)	G, soft, coagulated at room temperature over 24 hours, cylindrical

(*continued*)

TABLE 16.2 (*continued*)

Name	Translation	Characteristics
Chaource	Name of town	C, coagulated over 12 hours, whey drained without pressing, eaten fresh or aged
Chevrotin	Goat milk	G, based on Reblochon, made in Upper Savoy
Comté	County	See Chapter 12
Crottin de Chavignol	Horse dung (referring to shape) of Chavignol (a town)	G, similar to Chabichou du Poitou but drum-shaped
Époisses de Bourgogne	Époisses (a town) of Burgundy (a region)	See Chapter 8
Fourme d'Ambert	Form (for shaping curd) of Ambert (a town)	C, semi-hard, injected with *Penicillium* spores and later with air, mild flavor
Fourme de Montbrison	Cheese of Montbrison (a town)	C, almost identical to Fourme d'Ambert
Laguiole	Name of town	C, milk obtained at 2600 feet or higher, pressed twice, thick rind with aluminum plaque for identification
Langres	Name of plateau	C, similar to Époisses de Bourgogne, sunken top, sometimes eaten with champagne poured on it
Livarot	Name of town	See Chapter 8
Maroilles (or Marolles)	Name of abbey in Ardennes forest	See Chapter 8
Mont d'Or (or Vacherin du Haut-Doubs)	Mont d'Or (a region in France and Switzerland) or young cowherd of Upper Doubs (an area)	C, smear-ripened, spruce wood used to hold wheels together, ripen upon, and package in
Morbier	Name of town	See Chapter 7
Munster or Munster-Géromé	Name of town	See Chapter 8

TABLE 16.2 (*continued*)

Name	Translation	Characteristics
Neufchâtel	Name of town	C, ripe cheese kneaded into fresh curd, appearance similar to Camembert
Ossau-Iraty	Names of valleys	S, lightly pressed, varies with cheesemaker
Pélardon	Small goat cheese	G, soft, uncooked, made in Languedoc-Roussillon region
Picodon	Spicy	G, similar to Pélardon, from Rhône river valley, six recognized sub-varieties
Pont-l-Évêque	Name of town	See Chapter 8
Pouligny-Saint-Pierre	Name of town	G, similar to Chabichou du Poitou but shaped like Eiffel Tower or pyramid
Reblochon de Savoie	Re-pinch the udder of Savoy	See Chapter 8
Rocamadour	Name of town	G, aged 12–15 days, small disks weighing 11/3 ounces
Roquefort	Name of town	See Chapter 9
Sainte-Maure de Touraine	Name of town	G, surface mold, log-shaped with stick or wheat straw through middle to cheese together and introduce oxygen
Saint-Nectaire	Marshal of Sennecterre, who introduced it (along with Cantal and Salers) to Louis XIV	C, smear-ripened, purple-gray rind with orange and white mold
Salers	Cattle breed	See Chapter 10
Selles-sur-Cher	Name of town	G, similar to Chabichou du Poitou but size and shape of hockey puck and coated with charcoal

(*continued*)

TABLE 16.2 (*continued*)

Name	Translation	Characteristics
Tome des Bauges	Alpine cheese of Bauges (mountains)	C, uncooked, pressed, molds on surface
Valençay	Name of town	G, similar to Chabichou du Poitou but shaped like truncated pyramid coated with ash

Germany

The German landscape is often hilly and mountainous, but it is not rough enough to preclude the breeding of dairy cattle. Allgäu is made in Bavaria, Altenburg comes from Saxony, and the Odenwald mountains are in Hesse. (See Table 16.3.)

Greece

Greek cheeses are predominately made from milk of goats and sheep, as the rugged stony terrain is not conducive to cattle breeding. Several whey cheeses are on the list. (See Table 16.4.)

Italy

Italy has more PDO cheese varieties than any country outside France. The country encompasses the Alps, river basins, plains, and the Mediterranean coast, resulting in a variety of landscapes, dairy species, and cheese types. Many Italian cheeses are hard, low-moisture varieties that can withstand extremes of temperature. (See Table 16.5.)

Netherlands

The Dutch make many varieties of cheese, but only six have the PDO designation to date. Dutch cheeses are made from cow's milk because the flat, rich soil is ideal for raising dairy cattle. Note that the last cheese is named "Noord-Hollandse

TABLE 16.3

German PDO cheeses

Name	*Translation*	*Characteristics*
Allgäuer Bergkäse	Mountain cheese from Allgäu cattle	C, similarities to Emmental, made in late spring and summer only
Allgäuer Emmentaler	Emmental from Allgäu cattle	C, very similar to Emmental, but smaller and faster ripening
Altenburger Ziegenkäse	Goat cheese from Altenburg (a town)	G, soft, can contain caraway seeds and up to 85% cow's milk, mold-dusted rind
Odenwälder Frühstückskäse	Breakfast cheese from Odenwald (mountains)	C, soft, small flat disks, washed rind

TABLE 16.4

Greek PDO cheeses

Name	*Translation*	*Characteristics*
Anevato	Anevato (a town)	G, sharp, salmon-pink whey cheese, from Grevena and Voios
Batzos	Hut used to milk animals	G or S, semi-hard pickled cheese from northern Greece, originally a byproduct of Manouri manufacture
Feta	Slice	See Chapter 5
Formaella Arachovas Parnassou	Cheese of Arachova, Parnassos	G or S, shaped in wicker cylinders and immersed in hot whey
Galotiri	Milk cheese	S, soft, hung until needed, from Epirus and Thessalia
Graviera Agrafon	Gruyère from Agrafa Mountains	G and/or S, sharp Gruyère type, dry-salted over 2–3 weeks
Graviera Kritis	Gruyère from Crete	G, similar to Graviera Agrafon

(*continued*)

TABLE 16.4 (*continued*)

Name	*Translation*	*Characteristics*
Graviera Naxou	Gruyère from Naxos	C, Graviera Agrafon but mild
Kalathaki Limnou	Little basket of Limnos (island)	G, similar to Feta, held in a wicker basket while the whey is drained
Kasseri	Kars (a town in Turkey)	S, stretched curd, from Lesvos and northwestern Greece
Katiki Domokou	Domokos (a district) goat	G, consistency of thick yogurt, eaten fresh and used as spread
Kefalograviera	Combination of Kefalotyri (another variety) and Graviera	S, cross between very hard Kefalotyri and Graviera, from northwestern Greece
Kopanisti	Beaten	C, S, or G, soft, aged with fresh cheese kneaded in, from Cyclades Islands
Ladotyri Mytilinis	Olive oil cheese from Mytilinis	S, hard, preserved in olive oil
Manouri	Cream cheese	G or S, soft whey cheese made throughout Greece with 25% added milk or cream
Metsovone	Combination of Metsovo (a town) and Provolone	Primarily C, similar to Provolone, smoked
Pichtogalo Chanion	Thick milk from Hania	Blend of G and S, consistency of thick yogurt, eaten fresh
San Michali	St. Michael	C, very hard, curd pressed four times, from Syros Island
Sfela	Slice	S, similar to Feta, curd reheated after cutting, from southern Peloponnese
Xigalo Siteias	Acid milk of Sitea region	G or S, uncooked and unstirred, curds separated after 30–45 days at room temperature
Xynomyzithra Kritis	Sour whey cheese of Crete	G or S, from whey with up to 15% whole milk, ripened in cloth bags

TABLE 16.5

Italian PDO cheeses

Name	*Translation*	*Characteristics*
Asiago d'Allevo	Raised Asiago (a plateau)	Part skim C, cooked twice, lightly pressed, aged 3 months to 2 years
Asiago Pressato	Pressed Asiago	Whole C, regular pressing, aged 1 month
Bitto	Bitto (a river)	See Chapter 13
Bra	Bra (a town)	Skim C, sometimes with added G or S, aged 45 days to 2 years
Caciocavallo Silano	Sila (a plateau) cheese on horseback	See Chapter 6
Canestrato Pugliese	Puglia (a region) woven basket	S, drained and shaped in basket, dipped in very hot water or whey to form rind, which is rubbed with olive oil
Casatella Trevigiana	Woman of the home in Treviso	C, soft, uncooked, eaten fresh
Casciotta d'Urbino	(Small circle of) cheese from Urbino (a town)	S with 30% C, aged a month, made December to September, comes in 2-pound wheels
Castelmagno	Castelmagno (a town)	Primarily C, aged in caves, often eaten before blue veins develop
Fiore Sardo	Flower of Sardinia	S, hard, uncooked, dipped in very hot water or whey to form rind, which is rubbed with olive oil, very strong flavor
Fontina	Fontin (a family)	Whole C, made in Aosta Valley as soon as cows are milked
Formai de Mut dell'Alta Valle Brembana	Alpine cheese of the high Brembana Valley	C, similar to Fontina but made from two milkings
Gorgonzola	Gorgonzola (a town)	See Chapter 9

(*continued*)

TABLE 16.5 (*continued*)

Name	*Translation*	*Characteristics*
Grana Padano	Grainy of Po (valley)	See Chapter 13
Montasio	Montasio (a mountain mass in Fruili)	Part skim C, similar to Asiago, fresh (2 months), medium (5–10 months), and aged (1–4 years) versions
Monte Veronese	Verona (a province) mountain	Whole C (similar to Asiago Pressado) or part skim C (aged 3 months to 3 years)
Mozzarella di Bufala Campana	Buffalo Mozzarella of Campania (a region)	See Chapter 6
Murazzano	Murazzano (a town)	S with up to 40% C, not cooked or pressed, eaten within 2 weeks
Nostrano Valtrompia	Our Trompia (a valley)	C, very hard, saffron added, aged at least a year
Parmigiano Reggiano	Parma and Reggio Emilia (provinces)	See Chapter 13
Pecorino di Filiano	Sheep cheese of Filiano (a town)	S, hard, surface rubbed with olive oil and vinegar, cave ripened
Pecorino Romano	Roman sheep cheese	S, very hard, coagulated with lamb rennet, wheels weigh 44–77 pounds
Pecorino Sardo	Sardinian sheep cheese	S but coagulated with calf rennet, semi-cooked, mild and matured versions
Pecorino Siciliano	Sicilian sheep cheese	S, drained in baskets that impart pattern on rind, must be aged 4 months
Pecorino Toscano	Tuscany sheep cheese	S but coagulated with calf rennet, semi-cooked, relatively soft
Provolone Valpadana	Little Provola of Po Valley	C, stretched curd, fresh version coagulated with calf rennet, piquant version uses kid or lamb rennet

TABLE 16.5 (*continued*)

Name	*Translation*	*Characteristics*
Quartirolo Lombardo	Fourth (referring to growth following third mowing) Lombardy	C, fresh version eaten within a month, more mature version develops red rind
Ragusano	Ragusa (a city)	C, similar to Caciocavallo but in blocks weighing 26–35 pounds, sometimes indented around middle from ropes used to hang it
Raschera	Rachera (a lake)	Part skim C, G, or S, springy texture, aged 1–6 months
Ricotta Romana	Roman recooked	See Chapter 5
Robiola di Roccaverano	Red of Roccaverano (a town)	Traditionally G and aged 3 days, may contain up to 85% C and 15% S and then aged 3 weeks, develops reddish rind, similar to Taleggio
Spressa delle Guidicarie	Pressed of Guidicarie (a town)	Part skim C, aged up to 6 months, texture changes from elastic to granular
Squacquerone di Romagna	Wobbly from Romagna (a region)	C, very soft and runny, 5-day shelf life
Stelvio (or Stilfser)	Stelvio-Stilfser (a national park)	C, pressed, smear ripened, aged 60 days
Taleggio	Taleggio (a valley)	C, smear ripened, relatively mild aroma
Toma Piemontese	Piedmont fall (referring to curds sinking in vat)	C, semi-hard, varies depending on cheesemaker
Valle d'Aosta Fromadzo	Aosta Valley part skim cheese	C, similar to Fontina but contains part skim milk from two milkings
Valtellina Casera	Valtellina (a region) aging cellar	C with up to 10% G, non-seasonal version of Bitto

Gouda" despite the fact that the town of Gouda is in southern Holland. (See Table 16.6.)

Portugal

Sandwiched between the Atlantic Ocean and the mountains, Portugal has not developed as extensive a cheese industry as neighboring Spain. Nevertheless, the country is well represented on the PDO list. Several Portuguese cheeses are coagulated with cardoon thistle (see Chapter 2). Queijo da Beira Baixa covers three varieties: Queijo Amarelo da Beira Baixa, Queijo de Castelo Branco, and Queijo Picante da Beira Baixa. (See Table 16.7.)

TABLE 16.6

Dutch PDO cheeses

Name	*Translation*	*Characteristics*
Boeren-Leidse met Sleutels	Leiden farmers with crossed keys	Low-fat, farm made, contains cumin seeds, Leiden's crossed keys coat of arms stamped on rind
Kanterkaas	Sharp edge cheese	Hard, sides of wheel are straight instead of bulging, originated in Friesland
Kanteromijnekaas	Sharp edge cumin cheese	Same as Kanterkaas but containing cumin seeds, originated in Friesland and Westerkwartier region
Kanternagelkaas	Sharp edge nail cheese (cloves resemble nails)	Same as Kanterkaas but containing cloves, originated in Friesland
Noord-Hollandse Edammer	North Holland Edam	Similar to Edam but less salty, skimmed evening milk mixed with whole morning milk, spherical
Noord-Hollandse Gouda	North Holland Gouda	Similar to Gouda but less salty

TABLE 16.7

Portuguese PDO cheeses

Name	*Translation*	*Characteristics*
Queijo Amarelo da Beira Baixa	Yellow cheese of Lower Beira region	G and S blend, flavor and texture vary widely, yellow rind
Queijo de Azeitão	Cheese of Azeitão region	S, coagulated with cardoon thistle, brine-washed, eaten by spooning out interior, from Serra da Arrábida area
Queijo de Cabra Transmontano	Goat cheese from beyond the mountains	G, very hard and rare, from Trás-os-Montes
Queijo de Castelo Branco	Castelo Branco (a town) cheese	S, coagulated with cardoon thistle, mild and crumbly
Queijo de Évora	Cheese of Évora (a town)	S, coagulated with cardoon thistle, aged over 60 days, hard and crumbly, many small eyes
Queijo de Nisa	Cheese of Nisa (a town)	S, similar to Queijo de Évora but semi-hard
Queijo do Pico	Cheese of Pico Island	C, semi-soft, aged over 20 days, intense aroma
Queijo Mestiço de Tolosa	Mixed cheese of Tolosa district	G and S blend, may be rubbed with ground paprika and water
Queijo Picante da Beira Baixa	Piquant cheese of Lower Beira region	G and S blend, gray rind, very strong spicy flavor
Queijo Rabaçal	Rabaçal parish cheese	Blend of 80% G and 20% S, coagulated with goat rennet, wheels are ½–1 pound
Quiejo São Jorge	St. George Island cheese	C, crumbly and similar to Cheddar, wheels weigh 18–26 pounds, rind may be mottled with red-brown mold
Quiejo Serpa	Serpa (a town) cheese	S, coagulated with cardoon thistle, rind has blue-gray mold, eaten by cutting off top and spooning out interior
Quiejo Serra da Estrela	Serra da Estrela ("Star Mountains") cheese	See Chapter 8
Queijo Terrincho	Terrincha sheep cheese	S, rare, 2- to 3-pound wheels tied with straw

Spain

Spain experiences a variety of climates: Alpine, Mediterranean, oceanic, semi-arid, and subtropical. The conditions for grazing animals therefore vary greatly, leading to a large range of cheeses. Afuega'l pitu, from the Asturian region, has been translated from the local language as "stick in the throat," "suffocate the chicken," "hard to swallow," or more literally "fire in the palate" or "sets fire to your gullet." The first three translations apparently come from ancient practice of testing the curds by feeding them to a chicken: they were drained enough if the bird had trouble swallowing them. The fire references come from the hot and spicy paprika version. (See Table 16.8.)

Switzerland

The Swiss landscape is hilly in the northern half of the country and mountainous in the south. Four-fifths of the cultivated land is used for grazing. Goats and sheep are usually not used for dairying, and all of their AOC cheeses are made from cow's milk. Tête de Moine derives its name either from the tax levy of one cheese per monk or its resemblance to a shaved monk's head after some of the cheese is scraped off. The scraping is performed by a pushing the cheese through a stainless steel rod that projects upward from the center of a round wooden board, and then turning a pivot arm with a knob on the top and a blade on the underside. The device, known as a *girolle*, was invented for Tête de Moine and may be applied to other varieties of similar shape and texture. (See Table 16.9.)

United Kingdom

England, Scotland, and Wales have temperate climates. Most of their cheese comes from cow's milk. Many varieties are not being produced in large quantities—of the 11 British PDO cheeses, three are made at only one place, and three others are not being made at all. (See Table 16.10.)

Other Countries

Belgium, Ireland, Poland, Slovakia, and Slovenia also make cheeses with the PDO appellation. (See Table 16.11.)

TABLE 16.8

Spanish PDO cheeses

Name	*Translation*	*Characteristics*
Afuega'l pitu	See above	C, made in northwest Spain, shaped like cone or pear, may be dyed with smoked paprika
Arzúa-Ulloa	Arzúa (a town) Ulloa (a river)	C, soft, almost runny interior, similar to Queso Tetilla but ripened at least 15 days
Cabrales	Name of town	Mix of C, G, and S, acquires mold from cave, blue-purple veins, often covered with sycamore leaves
Cebreiro	Name of town	S, soft, shaped like mushroom
Flor de Guîa	Thistle flower	C, S, made in Gran Canaria, Canary Islands, soft, coagulated with cardoon thistle, rare and made only by women
Gamonéu (or Gamonedo)	Name of town (Gamonedo is Spanish translation)	Mix of C, G, and S, similar to Cabrales, but firmer and smoked
Idiazábal	Name of town	S with at least 6% fat, lamb rennet used, aged in cave 2 months, then usually smoked in beechwood kiln and aged 1 year
Máhon-Menorca	Máhon, island of Minorca	See Chapter 13
Picón Bejes-Tresviso	Bejes and Tresviso, in Picos de Europa mountains	Mix of C, G, and S, similar to Cabrales but also smeared with *Brevibacterium linens*
Queso de la Serena	Cheese of La Serena region	S, coagulated with vegetable rennet, eaten by cutting off top and spooning out soft interior
Queso de l'Alt Urgell y la Cerdanya	Cheese of Alt Urgell and Cerdanya regions	C, semi-soft, aromatic bark placed on surface

(*continued*)

TABLE 16.8 (*continued*)

Name	*Translation*	*Characteristics*
Queso de Murcia	Cheese of Murcia region	G, pressed in grass-lined forms, eaten fresh
Queso de Murcia al Vino	Cheese of Murcia region with wine	G, same as Queso de Murcia but washed with local red wines
Queso de Valdeón	Cheese of Valdeón valley	C, similar to Cabrales, wrapped in sycamore leaves
Queso Ibores	Cheese of Ibores (a district)	G, semi-soft, rind may be coated with olive oil or smoked paprika
Queso Majorero	Cheese of Maxorata (old name for Fuerteventura Island)	G, lightly pressed in palm leaf forms, may be rubbed with oil, paprika, or roasted corn meal
Queso Manchego	Cheese of La Mancha (a region)	S, soft, crumbly, or very hard depending on age (2 months to 2 years), zigzag markings on side, surface may have dark green mold
Queso Nata de Cantabria	Cream cheese of Cantabria region	C, soft, eaten fresh
Queso Palmero	Cheese of La Palma, anary Islands	G, coagulated with goat rennet, lightly smoked or rubbed with oil, crumbly
Queso Tetilla	Small breast cheese	C, made in Galicia, shaped like a teat, eaten fresh
Queso Zamorano	Cheese of Zamora province	Churra S, hard and aged, intense and nutty flavor
Quesucos de Liébana	Small cheeses from Liébana region	Primarily C, may be smoked, 1-pound wheels
Roncal	Name of valley	S, hard and granular, often smoked, imprinted with cloth it is aged in
San Simón da Costa	Saint Simon of Costa	C, made in Galicia, shaped like teardrop, smoked over birch wood
Torta del Casar	Torte of Casar de Cáceres	S, similar to Queso de la Serena but coagulated with cardoon thistle

TABLE 16.9

Swiss AOC cheeses

Name	*Translation*	*Characteristics*
Berner Alpkäse	Bernese Alp cheese	Hard, washed with brine, aged 6–12 months
Berner Hobelkäse	Bernese planing cheese (referring to shaving off slices)	Same as Berner Alpkäse but aged 18 months and very hard
Emmentaler	Emmen Valley	See Chapter 12
Formaggio d'Alpe Ticinese	Cheese of Ticino Alps	May contain up to 30% G, semi-hard, aged on wood planks for at least 45 days
Gruyère	Name of town	See Chapter 12
L'Étivaz	Name of town	Made between May 10 and October 10 in Vaud canton, cooked in copper kettle over open wood fire
Raclette du Valais	Raclette (scrape) of Valais (a canton)	Smaller version of Gruyère, commonly heated with melted layer scraped onto food
Sbrinz	Brienz (a town)	See Chapter 13
Tête de Moine	See above	Made in Bern, similar to Gruyère but aged 3–6 months, 1½ pounds
Vacherin Fribourgeois	Young cowherd of Fribourg (a canton)	Uncooked, lightly pressed, similar to Fontina
Vacherin Mont d'Or	Young cowherd of Mont d'Or (a region in France and Switzerland)	Smear ripened, wheels held together by strip of bark, aged on wood plank and washed in brine

TABLE 16.10

British PDO cheeses

Name	*Characteristics*
Beacon Hill Traditional Lancashire Cheese	C, made on farm with blend of curds from 2–3 days of milking, cheddared, creamy and crumbly versions, made by one company
Blue Stilton Cheese	See Chapter 9
Bonchester	C, soft, similar to Camembert, only cheese on list originating in Scotland, revived in 1980s but not currently made
Buxton Blue	C, harder version of Blue Stilton but with annatto on outside, not currently made
Dovedale Cheese	C, interior mold, brined instead of dry salted, aged 3–4 weeks, not currently made
Single Gloucester	See Chapter 10
Staffordshire Cheese	C, wrapped in cloth, similar to Cheshire, revived in 1980s, made by one company
Swaledale Cheese	C, hard, covered with wax or blue-gray mold, revived in 1980s and made by one company
Swaledale Ewes Cheese	S, but otherwise same as Swaledale
West Country Farmhouse Cheddar	C, same as Cheddar but must be handmade on farm
White Stilton Cheese	C, same as Blue Stilton but without *Penicillium* or resulting veins

PGI Cheeses

These cheeses are only partially produced in their traditional geographic areas. (See Table 16.12.)

TSG Cheeses

The TSG label indicates that the cheese is manufactured in a traditional manner with no reference to geographic origin. (See Table 16.13.)

Some places obviously have a huge cheesemaking tradition. Box 16.3 shows how much cheese is eaten in various countries.

TABLE 16.11

Other PDO cheeses

Name	*Translation*	*Characteristics*
Bovški Sir	Bovec (a Slovenian sheep breed) cheese	S with up to 20% C or G, intense aroma and flavor, brittle but not crumbly
Bryndza Podhalanska	Bryndza from Podhale region of Poland	See Chapter 5
Fromage de Herve	Cheese from Herve, Belgium	C, hard, comes in sweet, spicy, and strong versions, depending on aging time
Imokilly Regato	Regato cheese from Imokilly, County Cork, Ireland	C, very hard, uses specially developed rennet, brined and dry-salted
Oscypek	Small javelin (in Polish)	S with up to 40% C, stretched curd, hand-shaped into double cone around a skewer, smoked
Slovenská Bryndza	Slovakian Bryndza	See Chapter 5

Other Special Cheeses

Plenty of other cheese varieties come from a specific location and are not on the above lists. Some have applied for an appellation and have not yet received it, others come from European countries outside the EU, and still others are relatively new products developed by a particular company. Many cheeses have originated in Mexico and Central America (such as the Oaxaca mentioned in Chapter 6), South America (Minas, which is aged in open air, is the major Brazilian variety) and Asia (including India's Paneer, covered in Chapter 4), but no appellation systems exist for those areas. A few cheeses would never be approved, such as Casu Marzu, which contains live flies (Box 16.4). In fact, cheese used to have mites crawling on it all time and a couple still do (Box 16.5). If you want to make cheese at home without things crawling on it, turn to the next chapter.

TABLE 16.12

PGI cheeses

Name	*Translation*	*Characteristics*
Danablu	Danish blue	See Chapter 9
Dorset Blue Cheese	From Dorset (England)	C, blue and green veins, also called Blue Vinney
Emmental Français Est-Central	Emmental from east central France	C, Similar to Emmental, 154-pound wheels
Esrom	From Esrom Abbey (Denmark)	C, similar to Port Salut, rubbed with brine over several months
Exmoor Blue Cheese	From Exmoor (England)	Jersey C, plant rennet used, blue and white mold cultures added
Queso de Valdeón	Cheese from Valdeón valley (Spain)	C, blue-veined interior mold
Queijo Mestiço de Tolosa	Mixed cheese of Tolosa (Portugal)	G and S mix, may be coagulated with vegetable rennet
Ser Korycinski Swojski	Homemade cheese of Korycin (Poland)	C, no processing except filtering and cooling, produced within 5 hours of milking, has eyes
Slovenský Oštiepok	Slovak smoked cheese	S with C added, hand-stretched, egg-shaped and decorated
Slovenská Parenica	Slovak steaming	S, fresh, stretched and wound into spirals, darkened by steaming
Svecia	Sweden (in Latin)	C, similar to Edam and Gouda, curds not cooked
Tomme de Pyrénées	Alpine cheese of French Pyrenees	C, G, or S, usually coated with black wax
Tomme de Savoie	Alpine cheese of Savoy (France)	See Chapter 12
Teviotdale	Teviotdale (on border between England and Scotland)	Full-fat Jersey C, brined instead of dry salted, not currently made
Wielkopolski ser Smažony	Greater Poland fried cheese	C, crumbled, aged on metal sheets for 2 days, mixed with butter and fried for 15 minutes

TABLE 16.13

TSG cheeses

Name	*Translation*	*Characteristics*
Boerenkaas	Farmer's cheese (in Dutch)	C, G, S, or water buffalo, similar to Gouda and Edam but must be made on farm; see Chapter 11
Hushållsost	Household cheese (in Swedish)	C, similar to Port Salut
Mozzarella	To cut off (in Italian)	See Chapter 6

BOX 16.3
CHEESE CONSUMPTION AROUND THE WORLD

On average, every man, woman, and child in America consumes over 32 pounds of cheese per year, and Canadians eat 27 pounds each. But people wolf down a lot more in some other countries. Here is a (Table B16.3) showing the figures for the 28 countries where recent statistics are available.

TABLE B16.3

Cheese consumption in various countries

Country	Consumption (pounds per capita)
NORTH AMERICA	
United States	32.6
Canada	27.1
Mexico	5.9
SOUTH AMERICA	
Argentina	24.9
Brazil	7.0
EUROPE	
Greece	68.4
France	57.4
Iceland	55.9

(*continued*)

BOX 16.3 (*continued*)

Country	Consumption (pounds per capita)
Germany	49.7
Switzerland	47.1
Netherlands	46.2
Italy	46.0
Finland	45.5
Sweden	41.6
Austria	38.3
Czech Republic	36.7
Norway	33.7
Hungary	24.2
United Kingdom	24.0
Poland	23.8
Slovakia	20.9
Spain	18.0
Russia	12.3
ELSEWHERE	
Turkey	42.7
Israel	36.1
Australia	26.4
New Zealand	13.2
Japan	3.7

BOX 16.4
DO YOU WANT FLIES WITH THAT?

"The whole round sea was one huge cheese, and those sharks the maggots in it."
—HERMAN MELVILLE, *Moby Dick*, 1851

"Maggot" is the common name for fly larva, including fly species that inhabit cheese. The most common fly associated with cheese is *Piophila casei*, also known as the "cheese skipper," so named because the larvae can propel themselves through the air. They feed

BOX 16.4 (*continued*)

on decomposing protein, which is why they are also found on dead animals, including humans. *Piophila casei* pupae (the stage between larvae and adult) were found on a packet of organs in the abdomen of an Egyptian mummy dating to approximately 170 BC. Their size and age are used to date corpses, but they prefer stored foods, especially cheese. Maggots, flies, and mites were commonly found on cheese until well into the twentieth century, but modern hygienic practices and storage systems have eliminated this annoyance.

One cheese variety deliberately includes flies and is illegal to sell, though it can be made for home consumption. Casu Marzu ("rotten cheese") is a sheep's milk cheese made in Sardinia and containing live larvae of *Piophila casei*. Pecorino is left in the open air with part of the rind removed to allow cheese skippers to lay eggs. When they hatch, the larvae eat their way into the cheese, breaking down the fat and protein to the point where some of the structure liquefies. The cheese is considered safe to eat as long as the maggots are alive. Cheese skippers are able to jump a few inches, so consumers are advised to protect their eyes. People who do not wish to eat live maggots will place Casu Marzu inside a paper bag and seal it. The maggots will bang around in the bag until they have suffocated, and the cheese is then consumed. *Guinness World Records 2009* listed it as the most dangerous cheese to human health. The Italian government considers it to be a traditional food that has been continuously produced for more than 25 years, making it exempt from some of their safety rules. This type of product will never fly in other countries, where regulations prohibit such a thing.

Cheese skipper.

BOX 16.5
MITES

"The cheese-mites asked how the cheese got there,
And warmly debated the matter;
The Orthodox said that it came from the air,
And the Heretics said from the platter.
They argued it long and they argued it strong,
And I hear they are arguing now;
But of all the choice spirits who lived in the cheese,
Not one of them thought of a cow."
—ARTHUR CONAN DOYLE, *Songs of Action*, 1898

Cheese mites include *Acarus siro* and *Tyrolichus casei*, and used to be considered inseparable from cheese. The 1811 dictionary cited in Box 1.2 lists "mite" as slang for cheesemonger. The very first science documentary movie was *Cheese Mites*, filmed through a microscope and shown in a theater in London's Leicester Square in 1903. The movie, which lasts less than a minute (and may be viewed online), showed mites of various sizes crawling around a piece of cheese. For some time thereafter, inexpensive microscopes often included packets of mites for the customers' viewing pleasure.

In the past, cheese mites were often consumed along with the cheese. In Thomas Reade's classic 1861 novel *The Cloister and the Hearth*, which was set in the fifteenth century, the protagonist was appalled to be served a cheese thoroughly consumed by mites, though his dining companions thought nothing of it ("These nauseous reptiles have made away with every bit of it." "Well, it is not gone far. By eating of the mites we eat the cheese to boot"). Even today, cheese mites are deliberately introduced into a couple of cheese varieties to produce a characteristic nutty and fruity flavor. When the cheese has aged sufficiently, it is covered with a powder of living and dead mites along with their feces and molted skins. Milbenkäse ("mite cheese") hails from Würchwitz, Germany, where a cheese mite statue stands. Caraway-flavored Quark is placed in a box containing *Tyrolichus casei*, and the product is consumed after aging. The rind is reddish brown after three months and black after a year. Mimolette Vieille (from the French *molle*, "soft," referring to the crust of the immature cheese, and *vielle*, "aged") originated in Lille, France as a version of Edam. It is shaped like a sphere and colored with annatto. *Acarus siro* are sprinkled on the surface, tuning the rind gray and crumbly after a year, making this variety resemble a rusty cannonball. If uncontrolled, *Acarus siro* can consume up to a quarter of a piece of cheese, leaving a powder nearly an inch high.

There are old village wives who can make
magnificent cheeses by rule of thumb or no
apparent rule while scientifically trained girls
freshly armed from college with thermometers,
percentages and other gadgets may fail
hopelessly—cheese-making is an art rather
than a science.

—JOHN SQUIRE, *Cheddar Gorge*, 1938

17

DO TRY THIS AT HOME

CHEESEMAKING MAY WELL have been art back in the 1930s, but the science has advanced to the point where we can predict what will happen most of the time. Not all of the time, though, for we are dealing with living things, bringing luck into the equation. Yes, some art is involved, since skill and experience are required to make a top-notch product, but even a beginner is capable of making cheese.

Equipment

You probably have some of the equipment you need in your kitchen: measuring cups, a whisk for stirring, a knife, and a stainless steel or ceramic pot with a lid. Everything coming in contact with milk should be stainless steel, glass, or enamel-lined because aluminum and iron will leach into the curds when milk is acidified, resulting in off-flavors. You may use plastic after the curds have formed. Finely woven cheesecloth will be necessary, but a white handkerchief or cotton dish towel may serve as a substitute. You must also have an accurate thermometer that reads to within a degree and goes up to 200°. If you are making hard cheese, you should create a cheese press. You can take a straight-sided plastic bucket or canister, punch out some small holes in the bottom, and position it elevated in a tray so the whey drips from the cheese. For smaller cheeses, you could use an arrangement such as a length of PVC pipe that fits into a PVC floor drain. Even a big metal can should be OK. The pressure is applied by placing bricks, hand weights, or a container of water atop the "follower," which is the name for the disk that sits on the cheese. You may have to fashion a follower that fits snugly in the press.

Unless you're making cream cheese, you will also need milk. If you have access to fresh milk from a farm, you're in good shape. Homogenized whole milk from the store has small fat globules that cling to the casein and don't undergo lipolysis readily, affecting the texture and flavor. Therefore, you may want to use skim milk and add cream to it before you begin. Other ingredients may include starter culture, rennet, calcium chloride ($CaCl_2$), and salt. The starter culture may be cultured buttermilk, which behaves like a mesophilic starter, or a spoonful of plain yogurt with live and active yogurt cultures, which would serve as a thermophilic starter. Rennet tablets are found in the pudding section of the supermarket, but you may want to send away for cheese rennet. Some calcium may be lost with pasteurization and homogenization, so if you are not using raw milk, the addition of $CaCl_2$ will make coagulation easier and the curd stronger. The powder may be familiar to you as a non-NaCl deicer on roads and as a water hardener in swimming pools, but you would want the food-grade version, which is also used to make pickles crisp. $CaCl_2$ may be purchased in places that sell home canning and brewing supplies. Use non-iodized salt such as kosher salt, as iodine inhibits the growth of starter culture. Chlorine would also kill the starter, so any water that is mixed with rennet or $CaCl_2$ must be non-chlorinated.

Most of all, you have to keep everything clean and sanitary. If starter bacteria are able to grow in cheese, so are pathogens and spoilage microbes. Everything gets sterilized before being used. Clean your pot with hot, soapy water; boil water in it; and immerse your utensils in the boiling water for five minutes. Then wash them off with diluted bleach (a tablespoonful in half a gallon of water) and rinse thoroughly. Wash off milk by first rinsing with cold water (hot water will cook the milk onto the pot or utensil). Don't forget to wash your hands and the work surfaces.

Making Soft Cheeses

The easiest cheese to make is Queso Blanco. Take a gallon of whole milk and heat it in a pot with occasional stirring (to prevent scorching) until the temperature reaches 180°. Then stir every 30 seconds for 5 minutes and remove from heat. Add 4–5 ounces of vinegar until curds form, and stir for 10 minutes. Line a colander with cheesecloth and pour the curds and whey in it. Then tie up the corners of the cloth and allow it hang over the sink until the dripping stops, which takes up to 3 hours. Add salt to taste, enclose the curd in plastic wrap, and you're done.

Cream cheese is also easy. Use a quart of light cream, add a teaspoon of cultured buttermilk, and allow the cream to coagulate at room temperature for 12 hours. Then pour, tie, and hang as above, but allow 12 hours for draining. Add salt, herbs, pieces of fruit or vegetable, etc., if desired, and place in a container to shape it, such as an empty butter tub.

Making Mozzarella

This variety will require rennet and a thermophilic starter. Warm 2 gallons of whole milk to 90° and dissolve a rennet tablet in ¼ cup of cold non-chlorinated water. Add 4 ounces of plain yogurt to the milk and stir for a minute. Dissolve ½ teaspoon $CaCl_2$ in ¼ cup of water and gently stir it in for a minute. Then add the dissolved rennet. Stir gently using an up and down motion so the rennet is thoroughly mixed, and allow the milk to rest, covered, for 45 minutes at 90°. If the curd is ready, you will be able to poke a finger in it and see the curd fall away to either side as you lift your finger. If you do not get this "clean break," wait some more, but do not stir the curd.

Take a knife and cut the curd into ¼-inch cubes. To do this, cut straight down along the opposite side of the pot and drag the knife toward you. Then cut parallel to the first cut and continue until you have sliced through all of the curd. Rotate the pot 90° and repeat. Cover and allow to rest for 15 minutes as the cubes sink in the whey and firm up. Pour into a colander lined with cheesecloth (you may want to save the whey for making Ricotta) and rinse the curds with cold water. Wrap up the curds with the cloth and hold overnight. Eat the rest of the yogurt, and clean and save the cup.

Next you will have to see if the curd, which is now in one solid mass, has reached the desired pH of 5.2–5.3. Heat a half gallon of water to 180°, drop in a forkful of curd, stir, and stretch the curd by pulling some of it with the fork while holding the rest of it against the side of the pot with the whisk. You should be able to pull the curd into a string because, at this pH level, enough of the CCP has been removed to make the curd stretchable. The curd should also have a glossy sheen as the casein becomes fully hydrated with water. If the curd strings and looks shiny, it is ready (if not, wait a couple more hours). Then cut the rest of the curd into ½-inch cubes and place it in the hot water. Once the curd is soft, fashion it into several balls and stretch each one in all directions for a minute. Shape the curd by stuffing it into the empty yogurt cup. Then remove it and float it in a gallon of cold non-chlorinated water to which a pound of salt has been added. When your Mozzarella has cooled, dry it off and wrap it.

To make Ricotta, take the fresh whey and heat it to 200°. Then add ¼ cup of vinegar to coagulate the whey protein. Strain through a fine cloth, and tie and hang. You may only get a cup of cheese per gallon of whey, but that's better than throwing the whey out.

Making Monterey Jack

Warm 2 gallons of whole milk to 88° and shut off the heat. Add mesophilic culture by sprinkling onto the milk. If you have sent away for a powdered culture, wait 5 minutes to allow it to be hydrated by the milk. Stir the culture using an up-and-down motion into the milk, and allow it to rest, covered, for 45 minutes. Dissolve ½ teaspoon $CaCl_2$ in a half a cup of water, and gently stir it in for a minute. Dissolve ½ teaspoon liquid rennet or ¼ of a rennet tablet in a half a cup of water. Add the rennet and stir gently up and down for a minute. Cover the pot, and allow the milk to coagulate at 88°F for 30 minutes. Check if the curds have a clean break; if not, wait another 10 minutes.

Once the curds have a clean break, cut them into ½-inch cubes and wait 5 minutes. Increase the heat slowly to 100° over a 30-minute period, stirring gently every minute to prevent the curds from sticking together. Hold at 100° for another 30 minutes, stirring gently every 2–3 minutes as the curds sink in the whey. Allow curds to rest for 5 minutes. Pour and scoop out the whey, leaving some around the curds, and stir at 100° for 15 minutes. Pour curds and whey into a colander lined with cheesecloth, and allow whey to drain. Transfer the curds and cloth into the pot, sprinkle in a tablespoon of salt, and gently mix by hand. Take the cloth full of curds, tie up the corners, flatten with your hands, and press for 6–8 hours with 10 pounds of weight. Remove the cheese from the cloth and rub a tablespoon of salt on the surface. Allow to air-dry for a day on a draining rack, flipping once after 12 hours. Wrap the cheese, or, if you prefer, obtain cheese wax and melt it over the entire surface. Age it at 55° for a month or two, flipping daily for consistent ripening.

Books describing artisanal cheesemaking are listed at the end of this chapter. You ought to consult these for the nuances of the procedures and for the many other varieties you can make at home.

Serving Cheese

Whether you call it a cheese board, cheese platter, or cheese tray, anybody can serve a selection of cheeses in a logical manner. The idea is to have different and contrasting classes represented, so serving Brie and Camembert together would not be optimal since they are similar. Samples of surface mold, interior mold, cheddared, Swiss-type, and very hard cheeses would work. A soft goat cheese would provide a contrast in texture, flavor, and species. Distinctive appearances are also appealing: the log of goat cheese, a round red Edam, a cheese with an interesting rind, and a wedge of another variety would look attractive on the plate. If you are looking for quality instead of a supermarket item, find a cheese shop, ask for recommendations, and taste samples.

Once you have the cheese, cut off the amount you will be serving and allow it to warm to room temperature. Many volatile compounds responsible for flavor are not as detectable under refrigeration, so leave the cheese on the counter or table, wrapped, for 30–60 minutes. Unwrapping cheese too early will cause drying out. Ideally, cheese is sliced with a sharp knife, although a wire slicer is also helpful. How do you cut a log of soft goat cheese, which falls apart easily? Slice it with dental floss. Have a separate knife for each cheese and serve on separate boards or plates to avoid intermingling. Make sure each cheese is identified.

Some people are a bit more formal with cheese tasting and prefer to go through a progression. Trying a strong cheese first will overpower the nuances of the milder-flavored examples. Instead, start with the milder cheeses and work up to blue-veined varieties. You may find yourself ordering them from "young" to "old," and with goat cheese first and blue cheese last.

We already covered wine and beer with cheese in Chapter 12. Although most people serve cheese with crackers, you should try crusty bread, olives, and nuts. The spread shown in Figure 17.1 has plates of each. Apples and pears are good accompaniments, but dried fruits are also fine. Some people feel that chocolate goes well with interior mold cheeses.

FIGURE 17.1 Cheese platters surrounded by bread, crackers, figs, nuts, olives, and chocolate. Each cheese is labeled and has its own knife.

Source: Author.

Allergies and reactions have to be considered. My wife is allergic to penicillin and therefore cannot eat blue-veined cheese, and some people can't stand the butyric acid aroma of varieties like Parmesan. If you have anybody with some sort of aversion to a type of cheese, you may have to avoid it or at least keep it segregated and covered. If you have any cheese left over, use a new wrapper for it because the old one won't reseal properly. If any mold starts growing on leftover cheese, you can simply cut off that section and enjoy the rest.

Speaking of leftovers, stale bits of cheese do not have to be discarded. In some countries, these pieces are mixed with raw milk to restart fermentation, and alcohol is later added to stop the process. The result is a pungent cheese spread, called Bruss in Italy and Fromage Fort ("strong cheese") in France. A Middle Eastern variety, Naboulsi, consists of a Feta-like cheese that is boiled in brine and flavored with spices.

Other Things to Try

We explained in Box 2.5 that cow's milk cheese is yellow because of the β-carotene in it, and milk from goats, sheep, and water buffalo does not have β-carotene, resulting in white cheese. But why is milk white to begin with? Let's begin with light. Light comes in the form of waves, as seen in Figure 17.2.

The distance between two successive wave crests (or two successive wave troughs) is the *wavelength*, which is measured in nanometers, abbreviated nm. There are 25,400,000 nm in an inch. Objects have colors because they reflect or scatter light of certain wavelengths and absorb light of other wavelengths. Violet light has a wavelength between 380 and 450 nm, blue is at 450–475 nm, green is at 495–570 nm, orange is at 590–620 nm, and red is at 620–750 nm. Grass is green because it contains chlorophyll, which absorbs blue and red light, and reflects other colors, mostly green. The chlorophyll in grass and other plants converts the blue and red light into energy.

If you search online for the question "Why is milk white?" you will encounter some answers that say "Because it contains casein, which is white." That is logical, but wrong. Casein micelles range from 30 to 680 nm in diameter, and the fat droplets in milk are not much larger, on the order of 1000–10000 nm. The tiny sizes of the micelles and fat droplets cause them to scatter light of all colors, creating the appearance of white light.

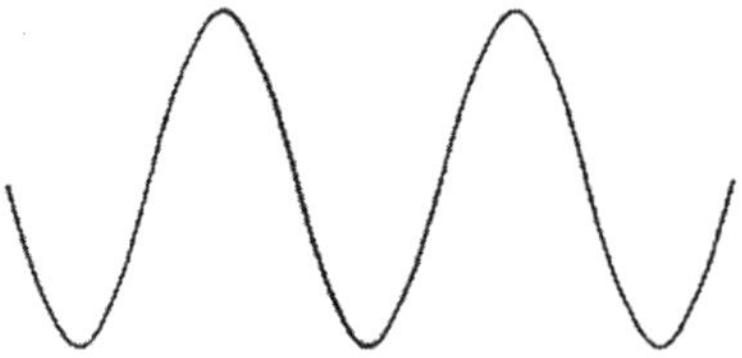

FIGURE 17.2 Waves of light.

The sky is blue for the same underlying reason, scattering of light. The nitrogen and oxygen atoms in the air reflect blue and violet the most. We see blue because our eyes are more sensitive to that color, and also because the sun radiates more blue light than violet. Sunlight passes through more air at sunrise and sunset, so there are more molecules to deflect blue and violet light. Those colors are scattered away, and red, orange, and yellow shine through.

Here's a dairy-related experiment illustrating this effect, and you can do it at home. Take a plain, clear drinking glass, fill it with water, and shine a flashlight through it. If you look down into the water, you will have difficulty seeing the beam. But if you add a few drops of milk, the beam becomes visible due to scattering by the micelles and fat droplets. If you add more milk, the color of the water will turn bluish, and if you look through the side of the glass opposite the flashlight, the beam will appear orange-red.

Another thing you can try at home is calculating the speed of light by heating cheese in a microwave oven. Like visible light, non-visible light is also in the form of waves. Infrared rays and microwaves have longer wavelengths than red light, and ultraviolet light and X-rays have shorter wavelengths than violet light. Infrared radiation is used for cooking because its energy level makes water molecules in food vibrate vigorously, creating friction and heat. Microwave ovens operate at a frequency of 2,450,000,000 hertz, meaning that the waves alternate from crest to trough almost 2½ billion times each second. The water molecules rotate as they try to align themselves with the alternating waves. As they rotate, they collide with other molecules, generating energy in the form of heat.

All light travels at the same speed, and you can find it by heating cheese in a microwave. First, cover a plate with slices of process cheese. Remove the turntable from the oven and replace it with the cheese plate. If necessary, elevate the plate so it will not turn when the oven is on. Cook on low heat until several melted spots appear (it takes about half a minute). Then remove the plate and measure the average distance in millimeters between the centers of the spots. That distance equals half the wavelength of the light, so double it and divide by 1000 to get the wavelength in meters. Finally, multiply by the frequency of 2,450,000,000 per second to get the speed of light in meters per second.

Science Fair Projects

Scientists learn about the universe by following the scientific method, which is laid out in the following steps:

1. Ask a question about something you are interested in.
2. Look up information to see what other scientists have done.

3. Come up with a hypothesis, which is an educated guess about how things work. You will have to predict what you think will happen in the experiment if your hypothesis is correct.
4. Conduct experiments to see if your hypothesis is right. Make sure you test everything more than once and that the results are repeatable.
5. Evaluate the results to explain the data, and draw a conclusion. If you see something unexpected, you may want to continue the research in a new direction.
6. Communicate your findings by presenting them at a meeting and writing a paper.

If you are a school student and if you apply the scientific method to some aspect of cheese, you should be able to discover things you did not know about before, and perhaps enter your project in a science fair. Even a student in elementary school is capable of performing experiments on cheese. The correct procedure is to vary just one thing at a time so you know what is causing the changes you observe. For instance, you can let some bread grow mold and then place the mold spores on different cheese samples. Then you can see if some cheese varieties grow more mold than others. You can also experiment with just one cheese variety and see if the growth of the mold differs with

FIGURE 17.3 Kira Pelletier and her winning science fair project. You can read about it at http://cheesemakinghelp.blogspot.com/2012/04/award-winning-science-project.html.

the type of wrapper (plastic, aluminum foil, wax paper, etc.). You could also make one variety of cheese, change one thing about the procedure, and see how it turns out. In 2012, a Pennsylvania sixth-grader named Kira Pelletier conducted experiments on which type of milk (raw, pasteurized, high-temperature pasteurized, powdered, evaporated) made the best Mozzarella cheese. She won second place in the biochemistry category at the Chester County Science Fair and advanced to the Delaware Valley Science Fairs, where she won a Future Scientist Award from our laboratory (Figure 17.3).

If you are more advanced, you could make a type of cheese with milk from cows, goats, and sheep, and mixtures of the milk, and investigate the differences between the cheeses. If you have access to a professional laboratory (and some colleges and industries will allow it, with supervision), you could see how the fatty acids or caseins change with age under various storage conditions. You could even try to make a new cheese variety. Remember that all of the cheese varieties in the world were developed as a result of experimentation and trial and error.

Books to Read

You should have no trouble finding an array of books about cooking and baking with cheese. Other books, many no longer in print, have been written about making, buying, serving, and enjoying cheese. Below is a list of ten non-technical books still available that cover these topics well. Some of these authors have written previous books that are also worth checking.

CHEESEMAKING AT HOME

Ricki Carroll, *Home Cheese Making: Recipes for 75 Homemade Cheeses* (Storey Publishing: North Adams, Mass., 2002). Originally titled *Cheesemaking Made Easy*, this popular book is in its third edition.

Jody M. Farnham and Marc Druart, *The Joy of Cheesemaking* (Skyhorse Publishing: New York, 2011). Describes cheesemaking in detail, along with stories of artisanal cheesemakers and recipes.

Mary Karlin and Ed Anderson, *Artisan Cheese Making at Home: Techniques and Recipes for Mastering World-Class Cheeses* (Ten Speed Press: Berkeley, Calif., 2011). The title says it all.

PURCHASING

Steve Jenkins, *Cheese Primer* (Workman Publishing, New York, 1996). Describes hundreds of cheeses from around the world, and gives highly opinionated advice about buying, storing, and serving them.

TASTING

Max McCalman and David Gibbons, *Mastering Cheese: Lessons for Connoisseurship from a Maître Fromager* (Clarkson Potter: New York, 2009). A wealth of information, focusing on how to be a connoisseur.

WINE AND CHEESE PAIRING

Laura Werlin, *All American Cheese and Wine Book: Pairing, Profiles, and Recipes* (Stewart, Tabori & Chang: New York, 2003). Won a James Beard Award for best single-subject book.

HISTORY

Paul S. Kindstedt, *Cheese and Culture: A History of Cheese and Its Place in Western Civilization* (Chelsea Green Publishing: White River Junction, Vt., 2012). The story of cheese from prehistoric times.

GENERAL INFORMATION

Juliet Harbutt, *World Cheese Book* (Dorling Kindersley: London, 2009). Pictures and descriptions of over 750 cheeses, with in depth discussions on the most important varieties.

Sharon Tyler Herbst and Ron Herbst, *The Cheese Lover's Companion: The Ultimate A-to-Z Cheese Guide with More Than 1,000 Listings for Cheeses and Cheese-Related Terms* (William Morrow: New York, 2007). A small encyclopedia of cheese.

Rob Kaufelt, *The Murray's Cheese Handbook: More Than 300 of the World's Best Cheeses* (Broadway Books: New York, 2006). About 300 varieties each get a paragraph, and there's a section with frequently asked questions.

OTHER RESOURCES

Two periodicals about cheese, *Culture Magazine* and *Cheese Connoisseur*, are published four times a year.

Arthur R. Hill's website at the Department of Food Science at the University of Guelph, Ontario, Canada, http://www.foodsci.uoguelph.ca/cheese/welcom.htm, contains technical information about all aspects of cheesemaking. And several blogs about cheese are on the Internet.

18

THE CHEESE STANDS ALONE

if a swiss cheese
could think
it would think that
a swiss cheese
was the most important
thing in the world
just as everything that
can think at all
does think about itself

—DON MARQUIS, *archygrams*, 1933

Cheese isn't the most important thing in the world, but it is one of the more pleasurable things life has to offer. What is nicer than being with friends, consuming and comparing various cheeses, accompanied by wine or beer? Making your own cheese is a rewarding experience, too. It may even lead to romance (see Chapter 28 of the Thomas Hardy novel *Tess of the D'Urbervilles*).

Cheese has been the subject of more than just the written word. Several countries have thought highly enough of cheese to have postage stamps (Figure 18.1) devoted to the varieties they have developed.

Mike Geno is an artist who creates oil-on-panel still life paintings of cheese, such as those below (Figure 18.2).

FIGURE 18.1 Postage stamps from France (left side), Portugal (top right, with a cardoon thistle postmark), Italy (right center), and Canada (bottom right).

FIGURE 18.2 A wall of Mike Geno's studio.
Source: Courtesy of Mike Geno.

FIGURE 18.3 "Kids and Calves" by Jim Victor.
Source: Courtesy of Jim Victor.

A neighbor of mine, Jim Victor, creates sculptures from food, including butter, chocolate, and cheese. He made the one shown in Figure 18.3 out of Cheddar for the Nebraska State Fair in 2008.

So why does cheese evoke enough passion to move people to create works of art? Mostly, I think, it is the multitude of aromas, flavors, and textures found in cheeses throughout the world, more than you encounter with any other food. It's not just food, it's a variety show. And all of the science behind the transformation from plants to milk to cheese is amazing. Cheese may be good and it may be bad, but the process behind its formation is never dull.

Cheese also provides nutrition in the form of vitamins, minerals, and high-quality protein. It comes in hundreds of varieties, and even within a variety, differences are noticeable: depending on origin of the milk, the microbes used, and the procedure the cheesemaker followed.

Our Fertile Crescent traveler from 8,000 years ago was certainly on to something.

Glossary

This shop is a dictionary; the language is the system of cheeses as a whole.
—ITALO CALVINO, *Mr. Palomar*, 1985

The following shows the definitions of terms that appear in more than one chapter. Words in italics refer to another term in the glossary. The chapter in which the term is most fully discussed is shown in parentheses at the end of the entry.

ACID A compound that dissolves in water and lowers its *pH* value. *Lactic acid* and vinegar are examples. (4)
ADJUNCT A bacterial culture added to milk to generate additional flavors during ripening. (2)
AFFINEUR A person in charge of aging cheeses, in a process called *affinage*. (3)
AGING Storage of cheese so that the proper flavor and texture develop. (3)
ALCOHOL A compound that has an *OH group*, where the oxygen atom is *bonded* to a carbon atom. (6)
ALDEHYDE A compound containing an oxygen atom that shares a *double bond* with the carbon atom at the end of the *carbon chain*. (10)
AMINO ACIDS The building blocks of proteins. Amino acids consist of an *amino group*, a *COOH group*, and a "side chain" containing carbon, hydrogen, and possibly nitrogen, oxygen, or sulfur. (5)
AMINO GROUP A part of a molecule where a nitrogen atom has two hydrogen atoms *bonded* to it. (5)
ANNATTO A plant-derived dye that imparts a yellow color to cheese. (2)

ARTISANAL CHEESE Hand-made cheeses made from local milk in small batches in a traditional manner. (1)

β-CAROTENE A red-orange compound that passes from cow's milk into cheese. It also acts as a vitamin A precursor. (2)

BOND A link between atoms. In food molecules, most of the bonds are single (the atoms share one electron) and some are *double bonds*. (1)

BRANCHED CHAIN A *carbon chain* with a fork in it. (5)

CALCIUM CHLORIDE ($CaCl_2$) A powder that is added to milk to bolster its calcium content, which aids coagulation and leads to a firmer *curd*. (2)

CALCIUM PHOSPHATE A compound containing calcium, phosphorus, and oxygen atoms. It is the principal form of calcium found in milk. Colloidal calcium phosphate (CCP) helps hold *micelles* together. (1)

CARBON CHAIN A string of carbon atoms *bonded* together. (1)

CASEIN The primary *protein* family in milk and cheese. It comes in four types: α_{s1}-casein, α_{s2}-casein, β-casein, and κ-casein. (1)

CHEDDARED CHEESE A cheese whose recipe includes flipping slabs of *curd* and stacking them to remove *whey* and allow acid development. Also known as English style cheese. (10)

CHEESE KNIVES Parallel wires strung across a metal frame, used to cut *curd* into cubes. (2)

COLLOIDAL Referring to miniscule particles suspended in a liquid. *Calcium phosphate* is colloidal in milk. (1)

CONJUGATED Alternating single *bonds* and *double bonds* in a *carbon chain*. Conjugated linoleic acid (CLA) appears to have health benefits. (5)

COOH GROUP A part of a molecule where a carbon atom (C) share *bonds* with two atoms of oxygen (O). One O is *double-bonded* to the C, and the other O is bonded to both the C and to a hydrogen atom (H). *Acids* found in cheese usually contain a COOH group. (1)

COOK Heating *curds* to facilitate bacterial growth and expulsion of *whey*. (2)

CURD The solid portion of milk after coagulation. (1)

DENATURATION When a *protein* unfolds and loses its biological activity. (1)

DOUBLE BOND A link between atoms in which they share two pairs of electrons. Double bonds are represented by the symbol = when depicting a molecular structure. (1)

ENZYME A special type of *protein* that speeds up chemical reactions. (1)

ESTER Fragrant compound where a carbon atom shares a *double bond* with an oxygen atom and also share a bond with (usually) another oxygen. (7)

EYES The round "holes" in cheese caused by the generation of carbon dioxide gas. Varieties with large eyes are also called Swiss type cheeses. (12)

FARMHOUSE CHEESE *Artisanal cheese* made on a farm from the milk of that farm. (1)

FAT A *triglyceride* that is solid at room temperature. (1)

FATTY ACID A *carbon chain* with a *COOH group* at one end; the other end attaches to a glycerol molecule. Fatty acids may be *saturated* or *unsaturated*. (1)

FLAVOR The combination of aroma, *taste*, and facial sensations (hot, cold, pungent, tingle, astringent). (7)

FRESH CHEESE A cheese that is not stored for more than a few weeks. (4)

FURAN Molecules containing one oxygen atom and four carbon atoms in a five-sided ring. (12)

GLOBULE A droplet. Fat in milk and cheese is in the form of globules. (1)

GREEN The aroma of cut grass. (4) Also, an unripened cheese. (2)

HOMOGENIZATION Conversion of fat *globules* to a small, uniform size by forcing milk through a narrow orifice under pressure. (2)

HYDROCARBON A compound containing only carbon and hydrogen atoms. (12)

INTERIOR-MOLD CHEESE A cheese which is pierced with needles to introduce oxygen and allow mold to grow inside. Also called blue cheese or blue-veined cheese, though the mold may be green. (9)

KETONE A carbon chain where one of the carbon atoms is linked to an oxygen atom with a *double bond.* (9)

LACTASE The enzyme (also known as β-galactosidase) that splits the *lactose* molecule into glucose and galactose. People who are able to digest lactose are lactase persistent. (1)

LACTIC ACID A product of the bacterial breakdown of *lactose.* (2)

LACTONE An *ester* where the portion containing the oxygen atoms is within a several-sided ring. (11)

LACTOSE The primary carbohydrate in dairy products. Milk is the only natural source. People who cannot digest lactose are lactose intolerant. (1)

LIPASE An *enzyme* that breaks down fat. (2)

LIPIDS *Triglycerides*, cholesterol, and the fat-soluble vitamins A, D, E, and K. (5)

LIPOLYSIS The breakdown of *lipids.* (5)

MESOPHILIC STARTER Bacteria that thrive best at 68°–104°. (2)

MICELLE A microscopic, spherical aggregate of *casein* molecules in milk. (1)

MOISTURE The amount of water in a food. (1)

OH GROUP A part of a molecule where an oxygen atom is *bonded* to a hydrogen atom and to a carbon atom. (6)

OIL A *triglyceride* that is liquid at room temperature. In this book, fats and oils are lumped together as fats. (1)

ORTHONASAL Perception of aroma by inhaling through the nose. (7)

PASTEURIZATION Heating to a temperature and length of time sufficient to kill pathogens. (2)

PEPTIDES Pieces of a *protein* molecule, consisting of some *amino acids* linked together. (5)

pH The measure of the acidity or alkalinity of a material. (2)

PICKLED CHEESE Cheese stored in a salt solution. Also called brined cheese. (5)

PROCESS CHEESE A blend of younger and older cheese. (14)

PROTEIN The three-dimensional structural network in cheese, consisting of *amino acids* connected together. Except for *whey cheeses*, nearly all of the protein in cheese is *casein.* (1)

PROTEOLYSIS The breakdown of *protein* into *peptides.* (5)

RAW MILK Milk that has not undergone *pasteurization.* (2)

RENNET Enzymes (mostly chymosin) present in the stomach of mammals, and also secreted by certain fungi and derived from some plants. A rennet preparation is added to milk to coagulate it into *curd.* (2)

RETRONASAL Perception of aroma by inhaling through the mouth. (7)

RIPENING Chemical and physical changes during *aging.* (3)

SATURATED A *fatty acid* containing no *double bonds* between carbon atoms. (5)

SCALDING Cutting *curd* into little pieces and heating it. (10)

SMEAR-RIPENED CHEESE A cheese that is washed with a yeast solution. Also called washed rind cheese (8).

SPECIES A group of similar living things capable of exchanging genes and interbreeding. (9)

STARTER CULTURE Bacteria that convert *lactose* to *lactic acid*, lowering the *pH* of milk at the start of cheesemaking. (2)

STRAIN Microorganism that is a variant of others within its *species*. (9)

STRETCHED CURD CHEESE A cheese whose *curd* is pulled like taffy before being placed in brine. Also called pasta filata cheese. (6)

SURFACE-MOLD CHEESE A cheese ripened from the outside in because of mold on the exterior. Also called bloomy rind cheese. (7)

TASTE Sensations of bitter, salty, sour, sweet, and umami as perceived by the taste buds. (7)

TERPENE A compound produced by pasture plants and containing isoprene, a structure with five carbon atoms, two *double bonds*, and a *branched chain*. (13)

TERROIR A French word describing the characteristics that altitude, soil, and weather give to a food. (13)

TEXTURE The perception of structural, surface, and mechanical properties of food. (10)

THERMOPHILIC STARTER Bacteria that thrive best at 86°–122°. (2)

***TRANS* FAT** A *fatty acid*, possibly harmful to health, in which *carbon chains* extend from opposite sides of a *double bond*. (5)

TRIGLYCERIDE Molecules consisting of three *fatty acids* attached to glycerol. The structure may be visualized as shaped like an "E." (1)

UNSATURATED A *fatty acid* containing at least one *double bond* between carbon atoms. (5)

WASHED-CURD CHEESE A cheese where the *curd* is rinsed by draining some of the *whey* and replacing it with hot water. (11)

WHEY The liquid portion of milk after coagulation. (1)

WHEY CHEESE A cheese made from coagulated *whey* proteins. (5)

References

"The Daily Telegraph heralded it as the most important cheese book ever written."
—GILES MORTON, *Edward Trencom's Nose: A Novel of History, Dark Intrigue, and Cheese*, 2008

The frequently cited references (preceded by their abbreviations in italics):

CCPM Fox, P.F.; McSweeney, P.L.H.; Cogan, T.M.; Guinee, T.P. (eds.). *Cheese: Chemistry, Physics and Microbiology*, 3rd ed. Elsevier Academic Press: San Diego, CA, 2004.

CLC Herbst, S.T.; Herbst, R. *The Cheese Lover's Companion: The Ultimate A-to-Z Cheese Guide with More Than 1,000 Listings for Cheeses and Cheese-Related Terms.* William Morrow: New York, 2007.

CP Jenkins, S. *Cheese Primer.* Workman Publishing: New York, 1996.

EDS Roginski, H.; Fuquay, J.W.; Fox, P.F. (eds.) *Encyclopedia of Dairy Sciences.* Academic Press: San Diego, CA, 2003

FCS Fox, P.F.; Guinee, T.P.; Cogan, T.M.; McSweeney, P.L.H. *Fundamentals of Cheese Science.* Aspen Publishers: Gaithersburg, MD, 2000.

500 Muir, R. *500 Cheeses.* Sellers Publishing: London, 2010.

WCB Harbutt, J. *World Cheese Book.* Dorling Kindersley Ltd.: New York, 2009.

USDA is the abbreviation for United States Department of Agriculture.

CHAPTER 1

General:

Kosikowski, F.V. Cheese. *Scientific American*, 1985, *252*(5), 88–99.
Salque, M.; Bogucki, P.I.; Pyzel, P.; Sobkowiak-Tabaka, I.; Grygiel, R.; Szmyt, M.; Evershed, R.P. Earliest evidence for cheese making in the sixth millennium BC in northern Europe. *Nature*, 2012, doi:10.1038/nature11698.

Mammals:

Galina, M.A.; Osnaya, F.; Cuchillo, H.M.; Haenlein, G.F.W. Cheese quality from milk of grazing or indoor fed Zebu cows and Alpine crossbred goats. *Small Ruminant Research*, 2007, *71*, 264–72.
Grandell, T. Whopping price for Swedish moose cheese at Europe's only moose dairy farm. Associated Press, June 23, 2004.
Jenness, R. In *Fundamentals of Dairy Chemistry*, 3rd ed.; N.P. Wong, ed. Van Nostrand Reinhold: New York, 1988, p. 21.
Or-Rashid, M.M.; Odongo, N.E.; Subedi, B.; Karki, P.; McBride, B.W. Fatty acid composition of yak (*Bos grunniens*) cheese including conjugated linoleic acid and *trans*-18:1 fatty acids. *Journal of Agricultural and Food Chemistry*, 2008, *56*, 1654–60.
Powell, R.L.; Norman, H.D. What breeds make up the national herd? *Hoard's Dairyman*, 2011, *156*(3):83.
Rath, S. *About Cows*. NorthWord Press: Minocqua, WI, 1987, pp. 17–24.
Thorpe, L. *The Cheese Chronicles*. HarperCollins: New York, 2009, pp. 357–64.
Wendorff, B.; Paulus, K. Impact of breed on the cheesemaking potential of milk; volume vs. content. *Dairy Pipeline*, 2011, *23*(1), 1, 4–7.

Making milk:

Bauman, D.E.; Mather, I.H.; Wall, R.J.; Lock, A.L. Major advances associated with the biosynthesis of milk. *Journal of Dairy Science* 2006, *89*, 1235–43.
Foley, R.C.; Bath, D.L.; Dickinson, F.N.; Tucker, H.A. *Dairy Cattle: Principles, Practices, Problems, Profits*. Lea & Febinger: Philadelphia, PA, 1972, pp. 372–80.

Casein micelles and fat globules:

FCS, pp. 25–39.
Kindstedt, P.S. *American Farmstead Cheese: The Complete Guide to Making and Selling Artisan Cheeses*. Chelsea Green Publishing: White River Junction, VT, 2012, pp. 45–46.
McGann, T.C.A.; Donnelly, W.J.; Kearney, R.D.; Buchheim, W. Composition and size distribution of bovine casein micelles. *Biochimica et Biophysica Acta*, 1980, *630*, 261–70.
Patton, S.; Keenan, T.W. The milk fat globule membrane. *Biochimica et Biophysica Acta* 1975, *415*, 273–309.

Other uses for milk:

Pearson, C.L. Animal glues and adhesives. In *Handbook of Adhesive Technology*, 2nd ed.; A. Pizzi, K.L. Mital, eds. Marcel Dekker: New York, 2003, pp. 479–94.
Sloane, E. *American Barns and Covered Bridges*. Funk & Wagnalls: New York, 1954, pp. 65–66.

Lactose intolerance and milk allergy:

Dunne, J.; Evershed, R.P.; Salque, M.; Cramp, L.; Bruni, S.; Ryan, K.; Biagetti, S.; di Lernia, S. First dairying in green Saharan Africa in the fifth millennium BC. *Nature* 2012, *486*, 390–94.

Guy, E.J.; Tamsma, A.; Knotson, A.; Holsinger, V.H. Lactase-treated milk provides base to develop products for lactose-intolerant populations. *Food Product Development* 1974, *8*(8), 50–74. Available at: http://wyndmoor.arserrc.gov/Page/1974\3943.pdf.

Itan, Y.; Jones, B.L.; Ingram, C.J.E.; Swallow, D.M.; Thomas, M.G. A worldwide correlation of lactase persistence phenotype and genotypes. *BMC Evolutionary Biology* 2010, *10*, 36–46.

Leonardi, M.; Gerbault, P.; Thomas, M.G.; Burger, J. The evolution of lactase persistence in Europe. A synthesis of archaeological and genetic evidence. *International Dairy Journal* 2012, *22*, 88–97.

Pelto, L.; Salminen, S.; Isolauri, E. Milk allergy. *EDS*, pp. 1821–28.

Etymology:

Barnhart, R.K. *The Barnhart Concise Dictionary of Etymology*. HarperCollins: New York, 1995.

Grose, F. *1811 Dictionary of the Vulgar Tongue: A Dictionary of Buckish Slang, University Wit, and Pickpocket Eloquence*. Digest Books: Chicago, 1971.

Hotten, J.C. *The Slang Dictionary*. 1913. Available at: http://openlibrary.org/books/OL7232088M/The_slang_dictionary.

Neilson, W.A. *Webster's New International Dictionary of the English Language*. G. & C. Merriam: Springfield, MA, 1940.

Onions, C.T. *The Oxford Dictionary of English Etymology*. Oxford University Press: Oxford, UK, 1982.

Teddington Cheese Wire web site. March/April 2000. Available at: http://www.teddingtoncheese.co.uk/acatalog/chwire/Issue10/cw10.htm.

American cows:

Blayney, D.P. *The Changing Landscape of U.S. Milk Production*. USDA Statistical Bulletin 978, June 2002. Available at: http://usda01.library.cornell.edu/usda/nass/SB978/sb978.pdf.

Hoard's Dairyman staff. Fewer dairy farms left the business. *Hoard's Dairyman* 2012, *157*, 147.

National Agricultural Statistics Service, Agricultural Statistics Board, USDA. *Milk Production*, April 19, 2012. Available at: http://usda01.library.cornell.edu/usda/nass/MilkProd//2010s/2012/MilkProd-02-17-2012.pdf.

Rathke, L. Small dairies go under as milk prices sink again. Associated Press, May 23, 2012.

Composition:

Calvo, M.S.; Whiting, S.J.; Barton, C.N. Vitamin D fortification in the United States and Canada: Current status and data needs. *American Journal of Clinical Nutrition* 2004, *80*(Suppl.), 1710S–1716S.

Kelly, A.L.; Fox, P.F. Indigenous enzymes in milk: A synopsis of future research requirements. *International Dairy Journal* 2006, *16*, 707–15.

Park, Y.W.; Juárez, M.; Ramos, M.; Haenlein, G.F.W. Physico-chemical characteristics of goat and sheep milk. *Small Ruminant Research*, 2007, *68*, 88–113.

USDA, Agricultural Research Service. USDA National Nutrient Database for Standard Reference, Release 23. Available at: http://www.ars.usda.gov/ba/bhnrc/ndl.

Milk from other species:

Mehaia, M.A. Fresh soft white cheese (Domiati-type) from camel milk: Composition, yield, and sensory evaluation. *Journal of Dairy Science* 1993, *76*, 2845–55.

Uniacke-Lowe, T. Studies on equine milk and comparative studies on equine and bovine milk systems. Ph.D. Thesis, University College Cork, 2011.

CHAPTER 2

General:

FCS, pp. 10–18.

Hill, A.H. University of Guelph Dept. of Food Science Cheese Site. Available at: http://www.foodsci.uoguelph.ca/cheese/welcom.htm.

Kosikowski, F.V.; Mistry, V.V. *Cheese and Fermented Milk Foods*, 2nd ed. F.V. Kosikowski LLC: Westport, CT, 1997.

McSweeney, P.L.H. *Cheese Problems Solved.* Woodhead Publishing: Cambridge, UK, 2007.

Ridgway, J. *The Cheese Companion*, 2nd ed. Running Press: Philadelphia, PA, 2004.

Starters:

Parente, E.; Cogan, T.M. Starter cultures: General aspects. *CCPM*, Vol. 1, pp. 123–47.

Wenner, M. Humans carry more bacterial cells than human ones. *Scientific American* online, November 30, 2007. Available at: http://www.scientificamerican.com/article.cfm?id=strange-but-true-humans-carry-more-bacterial-cells-than-human-ones.

Shape:

Appropriate sections of *CLC, CP, 500*, and *WCB*.

Kindstedt, P.S. *Cheese and Culture: A History of Cheese and Its Place in Western Civilization.* Chelsea Green Publishing: White River Junction, VT, 2012, pp. 181–83.

McNaughton, N. The shape can influence quality. *Cheese Reporter* 2002, *127*(8), 4.

Wason, B. *Encyclopedia of Cheese and Cheese Cookery*. Galahad Books: New York, 1966.

Milk standards:

Food and Drug Administration, U.S. Department of Health and Human Services. *Grade "A" Pasteurized Milk Ordinance*. 2009 Revision. Available at: http://www.fda.gov/downloads/Food/FoodSafety/ProductSpecificInformation/MilkSafety/NationalConferenceonInterstateMilkShipmentsNCIMSModelDocuments/UCM209789.pdf.

USDA, Agricultural Marketing Service, Dairy Programs. *Milk for Manufacturing Purposes and Its Production and Processing*. July 21, 2011 edition. Available at: http://www.ams.usda.gov/AMSv1.0/getfile?dDocName=STELDEV3004791.

Raw milk cheese:

Colonna, A.; Durham, C.; Meunier-Goddik, L. Factors affecting consumers' preferences for and purchasing decisions regarding pasteurized and raw milk specialty cheeses. *Journal of Dairy Science* 2011, *94*, 5217–26.

Latorre, A.A.; Pradhan, A.K.; Van Kessel, J.A.S.; Karns, J.S.; Boor, K.J.; Rice, D.H.; Mangione, K.J.; Gröhn, Y.T.; Schukken, Y.H. Quantitative risk assessment of listeriosis due to consumption of raw milk. *Journal of Food Protection* 2011, *74*, 1268–81.

National Association of State Departments of Agriculture. NASDA releases raw milk survey. Press release, July 19, 2011.

Van Hekken, D.L. Quality aspects of raw milk cheeses. *Food Technology* 2012, *66*(6): 67–78.

Why is cheese yellow?

Calderón, F.; Chauveau-Duriot, B.; Martin, B.; Graulet, B.; Doreau, M.; Nozière, P. Variations in carotenoids, vitamins A and E, and color in cow's plasma and milk during late pregnancy and the first three months of lactation. *Journal of Dairy Science* 2007, *90*, 2335–46.

Garber, L.L., Jr.; Hyatt, E.M.; Starr, R.G., Jr. Measuring consumer response to food products. *Food Quality and Preference* 2003, *14*, 3–15.

Lactic acid:

Kemp, G. Lactate accumulation, proton buffering, and pH change in ischemically exercising muscle. *American Journal of Physiology—Regulatory, Integrative and Comparative Physiology* 2005, *289*, R895–R901.

Lindinger, M.I.; Kowalchuk, J.M.; Heigenhauser, G.J.F. Applying physicochemical principles to skeletal muscle acid-base status. *American Journal of Physiology—Regulatory, Integrative and Comparative Physiology* 2005, *289*, R891–R894.

NSLAB:

Banks, J.M.; Williams, A.G. The role of the nonstarter lactic acid bacteria in Cheddar cheese ripening. *International Journal of Dairy Technology* 2004, *57*, 145–52.

Marquis, V.; Haskell, P. *The Cheese Book.* Simon and Schuster: New York, 1965; pp. 22–24.

Green cheese and green meat:

Wendorff, B.; Houck, K. Could cheese be the culprit when meat turns green? *Dairy Pipeline*, 2011, *23*(2), 1–2, 6–8.

Coagulation:

Lucey, J.A. Rennet coagulation of milk. *EDS*, pp. 286–88.

Whey utilization:

Farrell, H.M., Jr.; Jimenez-Flores, R.; Bleck, G.T.; Brown, E.M.; Butler, J.E.; Creamer, L.K.; Hicks, C.L.; Hollar, C.M.; Ng-Kwai-Hang, K.F.; Swaisgood, H.E. Nomenclature of the proteins of cows' milk-sixth revision. *Journal of Dairy Science* 2004, *87*, 1641–74.

Tunick, M.H. Whey protein production and utilization: A brief history. In *Whey Processing, Functionality and Health Benefits*; C.I. Onwulata, P. Huth, eds. Wiley & Sons: New York, 2008, pp. 1–13.

Giant cheeses:

DeWitt, D. *The Founding Foodies*. Sourcebooks: Naperville, IL, 2010, pp. 153–54.
Trout, G.M. Mammoth cheese. *Journal of Dairy Science* 1960, *43*, 1871–77.

CHAPTER 3

Affinage:

Gordinier, J. Cheese: A coming-of-age story. *New York Times*, October 5, 2011, p. D1. Available at: http://www.nytimes.com/2011/10/05/dining/cheese-and-affinage-a-coming-of-age-story.html?pagewanted=all.

Terroir:

Aurier, P.; Fort, F.; Sirieix, L. Exploring *terroir* product meanings for the consumer. *Anthropology of Food* 2005, *4*, at http://aof.revues.org/index187.html.

Length of time:

Stump, S. Say cheese! 40-year-old cheddar up for sale in Wisconsin. *Bites on Today* web site, August 30, 2012. Available at: http://bites.today.com/_news/2012/08/30/13573570-say-cheese-40-year-old-cheddar-up-for-sale-in-wisconsin?lite.

Processing:

Chambers, D.H.; Esteve, E.; Retiveau, A. Effect of milk pasteurization on flavor properties of seven commercially available French cheese types. *Journal of Sensory Studies* 2010, *25*, 494–511.
Hickey, D.K.; Kilcawley, K.N.; Beresford, T.P.; Wilkinson, M.G. Lipolysis in Cheddar cheese made from raw, thermized, and pasteurized milks. *Journal of Dairy Science* 2007, *90*, 47–56.

Flavor compounds:

Frank, D.C.; Owen, C.M.; Patterson, J. Solid phase microextraction (SPME) combined with gas-chromatography and olfactometry-mass spectrometry for characterization of cheese aroma compounds. *Lebensmittel-Wissenschaft und Technologie* 2004, *37*, 139–54.

Human milk:

Farrell, H.M., Jr.; Douglas, F.W., Jr. Effects of ultra-high-temperature pasteurization on the functional and nutritional properties of milk proteins. *Kieler Milchwirtschaftliche Forschungsberichte*, 1983, *35*, 345–56. Available at: http://wyndmoor.arserrc.gov/Page/1983\4781.pdf.
Kunz, C.; Lönnerdal, B. Re-evaluation of the whey protein/casein ratio of human milk. *Acta Paediatrica* 1992, *81*, 107–12.
Root, J. Mommy's Milk Cheese. *HowStuffWorks* web site. Available at: http://recipes.howstuffworks.com/mommys-milk-cheese-recipe.htm.

Winter, M. N.Y. chef keeps abreast of food trends with mother's-milk cheese. *USA Today*, March 9, 2010. Available at: http://content.usatoday.com/communities/ondeadline/post/2010/03/ny-chef-keeps-abreast-of-food-trends-with-mothers-milk-cheese/1#.T7unXFKwXEs.

What can go wrong:

Anonymous. Cheese producing region of Italy hit with second quake. *Cheese Market News*, June 1, 2012, pp. 1, 11.

Birkkjaer, H.E.; Soerensen, E.J.; Joergensen, J.; Sigersted, E. The influence of the cheesemaking technique on the quality of cheese. Danish Government Research Institute for Dairy Industry, Report 128, 1961.

CHAPTER 4

Cottage cheese:

Farkye, N.Y. Acid- and acid/rennet-curd cheeses. Part B: Cottage cheese. *CCPM*, Vol. 1, pp. 329–41.

Quark:

FCS, pp. 379–81.

Cream cheese:

Schulz-Collins, D.; Senge, B. Acid- and acid/rennet-curd cheeses. Part A: Quark, cream cheese and related varieties. *CCPM*, Vol. 1, pp. 301–27.

European varieties:

CP, pp. 175–77.
WCB, pp. 210–11.
CLC, pp. 79–80, 108.

Queso Blanco and Queso Fresco:

Van Hekken, D.L.; Farkye, N.Y. Hispanic cheeses: The quest for Queso. *Food Technology* 2003, *57*(1), 32–38. Available at: http://wyndmoor.arserrc.gov/Page/2002\7212.pdf.

Inspections:

USDA, Dairy Grading Branch. *DA Instruction 918-PS. Instructions for Dairy Plant Surveys*. March 3, 2008 edition. Available at: http://www.ams.usda.gov/AMSv1.0/getfile?dDocName=STELPRD3641026.

Pathogens:

Altekruise, S.F.; Timbo, B.B.; Mowbray, J.C.; Bean, N.H.; Potter, M.E. Cheese-associated outbreaks of human illness in the United States, 1973 to 1992: Sanitary manufacturing practices protect consumers. *Journal of Food Protection* 1998, *61*, 1405–7.

Bishop, J.R.; Smukowski, M. Storage temperatures necessary to maintain food safety. *Food Protection Trends* 2006, *26*, 714–24.

Brooks, J.C.; Martinez, B.; Stratton, J.; Bianchini, A.; Krokstrom, R.; Hutkins, R. Survey of raw milk cheeses for microbiological quality and prevalence of foodborne pathogens. *Food Microbiology* 2012, *31*, 154–58.

MacDonald, P.D.M.; Whitwam, R.E.; Boggs, J.D.; MacCormack, J.N.; Anderson, K.L.; Reardon, J.W.; Saah, J.R.; Graves, L.M.; Hunter, S.B.; Sobel, J. Outbreak of listeriosis among Mexican immigrants as a result of consumption of illicitly produced Mexican-style cheese. *Clinical Infectious Diseases* 2005, *40*, 677–82.

Ryser, E.T. Incidence and behavior of *Listeria monocytogenes* in cheese and other fermented dairy products. In *Listeria, Listeriosis, and Food Safety*, 2nd ed.; E.T. Ryser, E.H. Marth, eds. CRC Press: Boca Raton, FL, 1999, pp. 405–502.

Uhlich, G.A.; Luchansky, J.B.; Tamplin, M.L.; Molina-Corral, F.J.; Anandan, S.; Porto-Fett, A.C.S. Effect of storage temperature on the growth of *Listeria monocytogenes* on Queso Blanco slices. *Journal of Food Safety* 2006, *26*, 202–14.

CHAPTER 5

Feta:

Anifantakis, E.M.; Moatsou, G. Feta and other Balkan cheeses. In *Brined Cheeses*; A.M. Tamine, ed. Blackwell Publishing: Ames, IA, 2006, pp. 43–76.

Kondyli, E.; Katsiari, M.C.; Masouras, T.; Voutsinas, L.P. Free fatty acids and volatile compounds of low-fat Feta-type cheese made with a commercial adjunct culture. *Food Chemistry* 2002, *79*, 199–205.

Bryndza:

Dusinšký, R.; Ebringer, L.; Mikulášová, M.; Belicová, A.; Krajčovič, J. Slovak Bryndza cheese. In *Cheese: Types, Nutrition and Composition*; R.D. Foster, ed. Nova Science Publishers: Hauppauge, NY, 2011, pp. 143–55.

Amino acids:

Glavin, D.P.; Dworkin, J.P. Enrichment of the amino acid L-isovaline by aqueous alteration on CI and CM meteorite parent bodies. *Proceedings of the National Academy of Science (USA)* 2009, *106*, 5487–92.

Kaufman, M. *First Contact: Scientific Breakthroughs in the Hunt for Life Beyond Earth*. Simon and Schuster: New York, 2011, pp. 70–74.

Fatty acids:

Ashworth, U.S.; Ramaiah, G.D.; Keyes, M.C. Species difference in the composition of milk with special reference to the northern fur seal. *Journal of Dairy Science* 1966, *49*, 1206–11.

Amino acid and fatty acid breakdown:

McSweeney, P.L.H. Biochemistry of cheese ripening. *International Journal of Dairy Technology* 2004, *57*, 127–44.

McSweeney, P.L.H.; Sousa, M.J. Biochemical pathways for the production of flavor compounds in cheeses during ripening: A review. *Lait* 2000, *80*, 293–324.

Woo, A.H.; Kollodge, S.; Lindsay, R.C. Quantification of major free fatty acids in several cheese varieties. *Journal of Dairy Science* 1984, *67*, 874–78.

Yvon, M.; Rijnen, L. Cheese flavour formation by amino acid catabolism. *International Dairy Journal* 2001, *11*, 185–201.

Salt:

FCS, pp. 153–68.

Guinee, T.P. Salting and the role of salt in cheese. *International Journal of Dairy Technology* 2004, *57*, 99–109.

Stinky plants:

Kite, G.C.; Hetterscheid, W.L.A.; Lewis, M.J.; Boyce, P.C.; Ollerton, J.; Cocklin, E.; Diaz, A.; Simmonds, M.S.J. Inflorescence odours and pollinators of *Arum* and *Amorphophallus* (Araceae). In *Reproductive Biology*; S.J. Owens, P.J. Rudall, eds. Royal Botanical Gardens: London, 1998, pp. 295–315.

CHAPTER 6

Mozzarella:

Kindstedt, P.S. Mozzarella cheese: 40 years of scientific advancement. *International Journal of Dairy Technology* 2004, *57*, 85–90.

Rankin, S.A.; Chen, C.M.; Sommer, D.; Esposito, A. Mozzarella and Scamorza cheese. In *Handbook of Food Science, Technology, and Engineering*, Vol. 4; Y.H. Hui, ed. CRC Press: Boca Raton, FL, 2006, pp. 150–1 to 150–13.

Tunick, M.H.; Mackey, K.L.; Smith, P.W.; Holsinger, V.H. Effects of composition and storage on the texture of Mozzarella cheese. *Netherlands Milk and Dairy Journal* 1991, *45*, 117–25. Available at: http://wyndmoor.arserrc.gov/Page/1991\5615.pdf.

Melting:

Johnson, M. The melt and stretch of cheese. *Dairy Pipeline*, 2011, *12*(1), 1–5.

Joshi, N.S.; Muthukumarappan, K.; Dave, R.I. Effect of calcium on microstructure and meltability on part skim Mozzarella cheese. *Journal of Dairy Science* 2004, *87*, 1975–85.

Kindstedt, P.S. Factors affecting the characteristics of unmelted and melted Mozzarella cheese. In *Chemistry of Structure-Function Relationships in Cheese*; E.L. Malin, M.H. Tunick, eds. Plenum: New York, 1995, pp. 27–41.

Lawrence, R.C.; Creamer, L.K.; Gilles, J. Texture development during ripening. *Journal of Dairy Science* 1987, *70*, 1748–60.

Tunick, M.H.; Malin, E.L.; Smith, P.W.; Holsinger, V.H. Effect of skim milk homogenization on proteolysis and rheology of Mozzarella cheese. *International Dairy Journal* 1995, *5*, 483–91. Available at: http://wyndmoor.arserrc.gov/Page/1995\6157.pdf

Production:

Gould, B. *Understanding Dairy Markets. Cheese Production.* Available at: http://future.aae.wisc.edu/tab/production.html#9.

U.S. Census Bureau. *2012 Statistical Abstract.* Available at: http://www.census.gov/compendia/statab/2012/tables/12s0008.pdf.

USDA, Agricultural Marketing Service. *Dairy Market News.* 2011, *78*, Report 38 (September 19–23).

Maillard reaction:

Corzo, N.; Villamiel, M.; Arias, M.; Jiménez-Pérez, S.; Morales, F.J. The Maillard reaction during the ripening of Manchego cheese. *Food Chemistry* 2000, *71*, 255–58.

CHAPTER 7

Camembert and Brie:

Boisard, P. *Camembert: A National Myth.* University of California Press: Berkeley, CA, 2003.

Dusting with ash:

500, pp. 140–41.
CLC, 248–49.

Sulfur compounds:

Molimard, P.; Spinnler, H.E. Review: Compounds involved in the flavor of surface mold-ripened cheeses: Origins and properties. *Journal of Dairy Science* 1996, *79*, 169–84.

Sourabié, A.M.; Spinnler, H.E.; Bonnarme, P.; Saint-Eve, A.; Landaud, S. Identification of a powerful aroma compound in Munster and Camembert cheeses: Ethyl 3-mercaptopropionate. *Journal of Agricultural and Food Chemistry* 2008, *56*, 4674–80.

Sourabié, A.M.; Spinnler, H.E.; Saint-Eve, A.; Bonnarme, P.; Landaud, S. Recent advances in volatile sulfur compounds in cheese: Thiols and thioesters. In *Volatile Sulfur Compounds in Food.* M.C. Qian, X. Fan, K. Mahattanatawee, eds. ACS Books: Washington, DC, 2011, pp. 120–35.

Spinnler, H.-E.; Gripon, J.-C. Surface mould-ripened cheeses. *CCPM*, Vol. 2, pp. 157–74.

Appearance:

Jack, F.R.; Piggott, J.R.; Paterson, A. Discrimination of texture and appearance in Cheddar cheese using consumer free-choice profiling. *Journal of Sensory Studies* 1993, *8*, 167–76.

Aroma:

Gierczynski, I.; Guichard, E.; Laboure, H. Aroma perception in dairy products: The roles of texture, aroma release and consumer physiology. A review. *Flavour and Fragrance Journal* 2011, *26*, 141–52.

Malnic, M.; Hirono, J.; Sato, T.; Buck, L. Combinatorial receptor codes for odors. *Cell* 1999, *96*, 713–23.

Turin, L. *The Secret of Scent*. Harper Perennial: New York, 2007.
Zozulya, S.; Echeverri, F.; Nguyen, T. The human olfactory receptor repertoire. *Genome Biology* 2001, *2*(6), 1–12.

Flavor:

Hofmann, T.; Ho, C.-T.; Pickenhagen, W. Challenges in taste research: Present knowledge and future implications. In *Challenges in Taste Chemistry and Biology*. T. Hofmann, C.-T. Ho, W. Pickenhagen, eds. ACS Books: Washington, DC, 2004, pp. 1–24.

Sound:

Tunick, M.H.; Onwulata, C.I.; Thomas, A.E.; Phillips, J.G.; Mukhopadhyay, S.; Sheen, S.; Liu, C.-K.; Latona, N.; Pimentel, M.R.; Cooke, P.H. Critical evaluation of crispy and crunchy textures: A review. *International Journal of Food Properties* 2013, *16*, 949–963.

Texture:

Rosenthal, A.J. *Food Texture: Measurement and Perception*. Aspen Publishers: Gaithersburg, MD, 1999, pp. 1–17.
Szczesniak, A.S. Texture is a sensory property. *Food Quality and Preference* 2002, *13*, 215–25.

Taste:

Laugerette, F.; Passilly-Degrace, P.; Patris, B.; Niot, I.; Febbraio, M.; Montmayeur, J.-P.; Besnard, P. CD36 involvement in orosensory detection of dietary lipids, spontaneous fat preference, and digestive secretions. *Journal of Clinical Investigation* 2005, *115*, 3177–84.
Glendinning, J.I. Is the bitter rejection response always adaptive? *Physiology and Behavior* 1994, *56*, 1217–27.
Homma, R.; Yamashita, H.; Funaki, J.; Ueda, R.; Sakurai, T.; Ishimaru, Y.; Abe, K.; Asakura, T. Identification of bitterness-masking compounds from cheese. *Journal of Agricultural and Food Chemistry* 2012, *60*, 4492–99.
Kirimura, J.; Shimizu, A.; Kimizuka, A.; Ninomiya, T.; Katsuya, N. Contribution of peptides and amino acids to the taste of foods. *Journal of Agricultural and Food Chemistry* 1969, *17*, 689–95.
Yamaguchi, S. Basic properties of umami and its effects on food flavor. *Food Reviews International* 1998, *14*, 139–76.

CHAPTER 8

Limburger:

Leclercq-Perlat, M.-N.; Corrieu, G.; Spinnler, H.-E. The color of *Brevibacterium linens* depends on the yeast used for cheese de-acidification. *Journal of Dairy Science* 2004, *87*, 1536–44.

Reblochon:

WCB, pp. 74–75.

Smearing with alcohol:

Harbutt, J. *Cheese*. Willow Creek Press: Minocqua, WI, 1999, pp. 78–79.

Trappist style:

Englemann, E.; Holler, P. *Gourmet's Guide to Cheese*. H.F. Ullmann: Königswinter, Germany, 2009, pp. 110–11, 131.

Other smear-ripened cheeses:

Munster, *CLC*, pp. 305–7. Teleme, *CLC*, pp. 448–49. Tilsit, *FCS*, p. 232.

Aroma:

McGee, H. *On Food and Cooking*. Scribner: New York, 2004, pp. 58–59.
Parliment, T.P.; Kolor, M.G.; Rizzo, D.J. Volatile components of Limburger cheese. *Journal of Agricultural and Food Chemistry* 1982, *30*, 1006–8.
Urbach, G. The flavour of milk and dairy products: II. Cheese: Contribution of volatile compounds. *International Journal of Dairy Technology* 1997, *50*, 79–89.

Esters:

Curioni, P.M.G.; Bosset, J.O. Key odorants in various cheese types as determined by gas chromatography-olfactometry. *International Dairy Journal* 2002, *12*, 959–84.
Thierry, A.; Maillard, M.-B.; Richoux, R.; Lortal, S. Ethyl ester formation is enhanced by ethanol addition in mini Swiss cheese with and without added propionibacteria. *Journal of Agricultural and Food Chemistry* 2006, *54*, 6819–24.

Limburger in popular culture:

Schwartz, B. The cheese that stands alone. *Lapham's Quarterly*, Summer 2011. Available at: http://www.laphamsquarterly.org/roundtable/roundtable/the-cheese-that-stands-alone.php.

Foot odor and malaria:

Knols, B.G.J.; van Loon, J.J.A.; Cork, A.; Robinson, R.D.; Adam, W.; Meijerink, J.; De Jong, R.; Takken, W. Behavioural and electrophysiological responses of the female malaria mosquito *Anopheles gambiae* (Diptera: Culicidae) to Limburger cheese volatiles. *Bulletin of Entomological Research* 1997, *87*, 151–59.
Owino, E.A. Sampling of *An. gambiae* s.s mosquitoes using Limburger cheese, heat and moisture as baits in a homemade trap. *BMC Research Notes* 2011, *4*, 284–88.

CHAPTER 9

General:

Cantor, M.D.; van den Tempel, T.; Hansen, T.K.; Ardö, Y. Blue cheese. *CCPM*, Vol. 2, pp. 175–98.

Roquefort:

CP, pp. 156–57.
FCS, pp. 417–18.

Gorgonzola:

WCB, pp. 110–11.
Helm-Ropelato, R. The birthplace of Gorgonzola. Maybe. *Christian Science Monitor*, February 2, 2005. Available at: http://www.csmonitor.com/2005/0202/p11s02-lifo.html.
CP, pp. 206–11.

Stilton:

Gkatzionis, K.; Linforth, R.S.T.; Dodd, C.E.R. Volatile profile of Stilton cheeses: Differences between zones within a cheese and dairies. *Food Chemistry* 2009, *113*, 506–12.
Linford, J. *Great British Cheeses*. DK Publishing: New York, 2008, pp. 196, 208–13.
Stilton Cheesemakers' Association web site. Stilton: Britain's Historic Blue. Available at: www.stiltoncheese.com.

Danablu:

Waagner Nielsen, E. Danish cheese varieties. In *Cheese: Chemistry, Physics and Microbiology. Vol. 1: General Aspects*; 2nd ed.; P.F. Fox, ed. Aspen Publishers: Gaithersburg, MD, 1999, pp. 248–54.

Maytag Blue:

Lane, C.B.; Hammer, B.W. Method of making blue-veined cheese. U.S. Patent 2,132,077.

Species:

Gori, K.; Cantor, M.D.; Jakobsen, M.; Jespersen, L. Production of bread, cheese and meat. In *The Mycota. Vol. 10: Industrial Applications*; 2nd ed.; M. Hofrichter, ed. Springer Verlag: Berlin, 2010, pp. 1–28.
Marcellino, N.; Beuvier, E.; Grappin, R.; Guéguen, M.; Benson, D.R. Diversity of *Geotrichum candidum* strains isolated from traditional cheesemaking fabrications in France. *Applied and Environmental Microbiology* 2001, *67*, 4752–59.
Teuber, M. Genus II. Lactococcus. In *Bergey's Manual of Systematic Bacteriology. Vol. 3: The Firmicutes*; 2nd ed.; P. De Vos, G. Garrity, D. Jones, N.R. Krieg, W. Ludwig, F.A. Rainey, K.-H. Schleifer, W.B. Whitman, eds. Springer Verlag: New York, 2009, pp. 711–22.

Penicillin:

Lax, E. *The Mold in Dr. Florey's Coat: The Story of the Penicillin Miracle*. Henry Holt: New York, 2004, pp. 16–23, 204.
Stodola, F.H. Penicillin: Breakthrough to the era of antibiotics. *Science for Better Living: The Yearbook of Agriculture*. U.S. Government Printing Office: Washington, DC, 1968, pp. 339–44.

Another use:

Gerber, L.C.; Koehler, F.M.; Grass, R.N.; Stark, W.J. Incorporating microorganisms into polymer layers provides bio-inspired functional living materials. *Proceedings of the National Academy of Science (USA)* 2012, *109*, 90–94.

CHAPTER 10

General:

Linford, J. *Great British Cheeses*. DK Publishing: New York, 2008.

Cheddar:

Banks, J.M. Cheddar-type cheeses. *EDS*, pp. 356–63.
FCS, pp. 397–400.

Cheshire:

Lawrence, R.C.; Gilles, J.; Creamer, L.K. The relationship between cheese texture and flavor. *New Zealand Journal of Dairy Science and Technology* 1983, *18*, 175–90.
Tunick, M.H.; Nolan, E.J.; Shieh, J.J.; Basch, J.J.; Thompson, M.P.; Maleeff, B.E.; Holsinger, V.H. Cheddar and Cheshire cheese rheology. *Journal of Dairy Science* 1990, *73*, 1671–75. Available at: http://wyndmoor.arserrc.gov/Page/1990\5519.pdf.

Red Leicester and Double Gloucester:

BBC News. Cheese rolling. May 10, 2005. Available at: http://www.bbc.co.uk/gloucestershire/content/articles/2005/05/10/cheese_rolling_feature.shtml.
FCS, p. 398–401.

Other British varieties:

CP, pp. 316–18.
FCS, p. 402.
WCB, pp. 200–201, 216–17.

Aldehydes:

Le Quéré, J.-L.; Molimard, P. Cheese flavor. *EDS*, pp. 330–40.

Texture:

Chen, J.; Stokes, J.R. Rheology and tribology: Two distinctive regimes of food texture sensation. *Trends in Food Science and Technology* 2012, *25*, 4–12.
Foegeding, E.A.; Brown, J.; Drake, M.A.; Daubert, C.R. Sensory and mechanical aspects of cheese texture. *International Dairy Journal* 2003, *13*, 585–91.
Lawrence, R.C.; Creamer, L.K.; Gilles, J. Texture development during cheese ripening. *Journal of Dairy Science* 1987, *70*, 1748–60.

Lucey, J.A.; Johnson, M.E.; Horne, D.S. Perspectives on the basis of the rheology and texture properties of cheese. *Journal of Dairy Science* 2003, *86*, 2725–43.
Szczesniak, A.S. Texture is a sensory property. *Food Quality and Preference* 2002, *13*, 215–25.

Cheddar Gorge:

British Geological Survey. Cheddar Gorge. 2011. Available at: http://www.bgs.ac.uk/mendips/localities/cheddar.html.
Lyall, S. Tracing your family tree to Cheddar Man's mum. *New York Times* March 24, 1997. Available at: http://www.nytimes.com/1997/03/24/world/tracing-your-family-tree-to-cheddar-man-s-mum.html?src=pm.

Meow:

Rutherfurd, S.M.; Kitson, T.M.; Woolhouse, A.D.; McGrath, M.C.; Hendriks, W.H. Felinine stability in the presence of selected urine compounds. *Amino Acids* 2007, *32*, 235–42.

CHAPTER 11

Gouda and Edam:

FCS, pp. 403–5.
500, p. 202–3.
Van den Berg, G. Dutch-type cheeses. *EDS*, pp. 371–78.

Colby and Monterey Jack:

FCS, pp. 402–3.
Moss, W. The "true" story of Monterey Jack cheese. Monterey County Historical Society. Available at: http://www.mchsmuseum.com/cheese.html.

Lactones:

Alewijn, M.; Smit, B.A.; Sliwinski, E.L.; Wouters, J.T.M. The formation mechanism of lactones in Gouda cheese. *International Dairy Journal* 2007, *17*, 59–66.

Fodder vs. forage:

Bugard, C.; Buchin, S.; Noël, Y.; Tessier, L.; Pochet, S.; Martin, B.; Chamba, J. Relationships between Abondance cheese texture, its composition and that of milk produced by cows grazing different types of pastures. *Lait* 2001, *81*, 593–607.
Chilliard, Y.; Ferlay, A.; Doreau, M. Effect of different types of forages, animal fat or marine oils in cow's diet on milk fat secretion and composition, especially conjugated linoleic acid (CLA) and polyunsaturated fatty acids. *Livestock Production Science* 2001, *70*, 31–48.
Povolo, M.; Contarini, G.; Mele, M.; Secchiari, P. Study on the influence of pasture on volatile fraction of ewes' dairy products by solid-phase microextraction and gas chromatography-mass spectrometry. *Journal of Dairy Science* 2007, *90*, 556–69.

Moio, L.; Dekimpe, J.; Etievant, P.; Addeo, F. Natural volatile compounds in the raw milks from different species. *Journal of Dairy Research* 1993, *60*, 199–213.

Species:

Berger, Y.; Mikolayunas, C.; Thomas, D. Dairy sheep fact sheet. University of Wisconsin-Madison. Available at: http://www.dbicusa.org/documents/Dairy Sheep Fact Sheet.pdf.
FCS, pp. 177–78.
Shafie, M.M. Environmental effects on water buffalo production. *World Animal Review* 1993, *77*, 21–25.
Tranel, L. Beginner dairy goat fact sheet. Iowa State University. Available at: http://www.extension.iastate.edu/NR/rdonlyres/B090C051-8602-4456-B3D6-1ED769C2D495/46819/DairyGOATSFactSheet06.pdf.

Cannonballs:

Anonymous. Curiosities of naval literature. *Colburn's United Service Magazine* 1866 (Part II), 538–53.
Mythbusters. Episode 128. Available at: http://mythbustersresults.com/greased-lightning.

Feed:

Fedele, V.; Rubino, R.; Claps, S.; Sepe, L.; Morone, G. Seasonal evolution of volatile compounds content and aromatic profile in milk and cheese from grazing goat. *Small Ruminant Research* 2005, *59*, 273–79.
Toso, B.; Procida, G.; Stefanon, B. Determination of volatile compounds in cows' milk using headspace GC-MS. *Journal of Dairy Research* 2002, *69*, 569–77.

Differences between species:

Moio, L.; Dekimpe, J.; Etievant, P.; Addeo, F. Natural volatile compounds in the raw milks from different species. *Journal of Dairy Research* 1993, *60*, 199–213.

Bovine somatotropin:

Bauman, D.E.; Collier, R.J. Use of bovine somatotropin in dairy production. Cornell University and University of Arizona. Available at: http://agribiotech.info/details/2010 rBST article for NABC_Bauman 09-15 Final 04.pdf.
Brennand, C.P.; Bagley, C.V. Bovine somatropin in milk. Utah State University. Available at: https://extension.usu.edu/files/publications/factsheet/FN-250_6.pdf.
U.S. Food and Drug Administration. Report on the Food and Drug Administration's Review of the Safety of Recombinant Bovine Somatotropin. Available at: http://www.fda.gov/AnimalVeterinary/SafetyHealth/ProductSafetyInformation/ucm130321.htm.

CHAPTER 12

General:

Bachmann, H.-P.; Bütikofer, U.; Isolini, D. Swiss-type cheese. *EDS*, pp. 363–71.
CLC, pp. 65–66.

Fröhlich-Wyder, M.T.; Bachmann, H.P. Swiss cheese. *CPS*, pp. 246–67.
Kindstedt, P.S. *Cheese and Culture: A History of Cheese and Its Place in Western Civilization.* Chelsea Green Publishing: White River Junction, VT, 2012, pp. 146–57.

Eye defects:

Cakir, E.; Clark, S. Swiss cheese and related products. In *The Sensory Evaluation of Dairy Products*; S. Clark, M. Costello, M.A. Drake, F. Bodyfelt, eds. Springer: New York, 2009, pp. 427–57.
Federal Register, January 23, 2001, pp. 7458–60. Available at: http://www.gpo.gov/fdsys/pkg/FR-2001-01-23/html/01-2017.htm.

Furans:

Preininger, M.; Warmke, R.; Grosch, W. Identification of the character impact flavour compounds of Swiss cheese by sensory studies of models. *Zeitschrift für Lebensmittel-Untersuchung und-Forschung* 1996, *202*, 30–34.

Hydrocarbons:

Povolo, M.; Pelizzola, P.; Lombardi, G.; Tava, A.; Contarini, G. Hydrocarbon and fatty acid composition of cheese as affected by the pasture vegetation type. *Journal of Agricultural and Food Chemistry* 2012, *60*, 299–308.

Food pairing:

Ahn, Y.-Y.; Ahnert, S.E.; Bagrow, J.P.; Barabási, A.-L. Flavor network and the principles of food pairing. *Scientific Reports* 2011, *1*, 196–202.
Chartier, F. *Taste Buds and Molecules*. John Wiley & Sons: Hoboken, NJ, 2012, pp. 159–67.
Drahl, C. Strange plate-fellows. *Chemical and Engineering News*, June 18, 2012, pp. 37–40.
Madrigal-Galan, B.; Heymann, H. Sensory effects of consuming cheese prior to evaluating red wine flavor. *American Journal of Enology and Viticulture* 2006, *57*, 12–22.
Porzio, M. Letter to the Editor: Strange plate-fellows. *Chemical and Engineering News*, August 13, 2012, p. 2.

Wine and cheese:

Lambert, P. *The Cheese Lover's Cookbook and Guide*. Simon and Schuster: New York, 2000, pp. 345–47.
McCalman, M.; Gibbons, D. *Mastering Cheese: Lessons for Connoisseurship from a Maître Fromager*. Clarkson Potter: New York, 2009, pp. 218–45.
Werlin, L. *The All American Cheese and Wine Book: Pairings, Profiles, and Recipes*. Stewart, Tabori & Chang: New York, pp. 62–83.
Wolf, C. *American Cheeses*. Simon & Schuster: New York, 2008, pp. 32–33.

CHAPTER 13

Parmesan:

Kennedy, D. Italian producers bank on cheese. BBC News, August 2009. Available at: http://news.bbc.co.uk/2/hi/europe/8228131.stm.

Parmigiano-Reggiano and Asiago:

FCS, pp. 392–96.

Other very hard cheeses:

Roncal, *CLC*, pp. 407–8; *FCS*, p. 401.
Mahón-Menorca, *WCB*, pp. 154–55.
Bitto, *CLC*, pp. 85–86; *CP*, p. 216.
Kefalotyri, *CLC*, pp. 259–60; *FCS*, p. 401.
Sapsago, *CLC*, p. 421; *CP*, p. 288.

Terpenes:

Favaro, G.; Magno, F.; Boaretto, A.; Bailoni, L.; Mantovani, R. Traceability of Asiago mountain cheese: A rapid, low-cost analytical procedure for its identification based on solid-phase microextraction. *Journal of Dairy Science* 2005, *88*, 3426–34.
Köksal, M.; Chou, W.K.W.; Cane, D.E.; Christianson, D.W. Structure of 2-methylisoborneol synthase from *Streptomyces coelicolor* and implications for the cyclization of a noncanonical C-methylated monoterpenoid substrate. *Biochemistry* 2012, *51*, 3011–20.
Marais, J. Terpenes in the aroma of grapes and wines: A review. *South African Journal of Enology and Viticulture* 1983, *4*, 49–58.

More on **terroir***:*

Buchin, S.; Martin, B.; Dupont, D.; Bornard, A.; Achilleos, C. Influence of the composition of Alpine highland pasture on the chemical, rheological and sensory properties of cheese. *Journal of Dairy Research* 1999, *66*, 579–88.
Carpino, S.; Mallia, S. La Terra, S.; Melilli, C.; Licitra, G.; Acree, T.E.; Barbano, D.M.; Van Soest; P.J. Composition and aroma compounds of Ragusano cheese: Native pasture and total mixed rations. *Journal of Dairy Science* 2004, *87*, 816–30.
Tunick, M.H.; Van Hekken, D.L.; Call, J.; Molina-Corral, F.J.; Gardea, A.A. Queso Chihuahua: Effects of seasonality of cheesemilk on rheology. *International Journal of Dairy Technology* 2007, *60*, 13–21. Available at: http://wyndmoor.arserrc.gov/Page/2007\7781.pdf.

Baby cheese:

Brand, J. *Observations on Popular Antiquities*; Vol. 2. F.C. and J. Rivington: London, 1813, pp. 6–7.
Davidson, A. *Oxford Companion to Food*. 2nd ed. Oxford University Press: New York, 2000, p. 679.
CP, pp. 279–81.

Suffolk cheese:

British Cheese Board. History of Cheshire cheese. Available at: http://www.britishcheese.com/cheshire/history_of_cheshire_cheese-14.
Dickens, C. Chronicles of English counties: Suffolk. *All the Year Round*, 1885, *35*, 422–23.
Macdonald, J. *Feeding Nelson's Navy: The True Story of Food at Sea in the Georgian Era*. U.S. Naval Institute Press: Annapolis, MD, 2006, pp. 30–33.

Location, location, location:

Jacobsen, R. *American Terroir*. Bloomsbury USA: New York, 2010, pp. 218–41.
Paxson, H. Locating value in artisan cheese: Reverse engineering *terroir* for new-world landscapes. *American Anthropologist* 2010, *112*, 444–57.

CHAPTER 14

Processed cheese:

U.S. FDA. Cheeses and Related Cheese Products. *Code of Federal Regulations* Title 21, Pt. 133.167–133.180, 2011.
FCS, pp. 429–51.

Nutrition, vitamins and minerals:

Ervin, R.B.; Wang, C.-Y.; Wright, J.D.; Kennedy-Stephenson, J. Dietary intake of selected minerals for the United States population: 1999–2000. *Advance Data from Vital Health and Statistics*, April 27, 2004.
Food and Nutrition Board, Institute of Medicine, National Academy of Sciences. Dietary Reference Intakes: Recommended Intakes for Individuals. 2010. Available at: http://www.iom.edu/Activities/Nutrition/SummaryDRIs/~/media/Files/Activity Files/Nutrition/DRIs/5_Summary Table Tables 1–4.pdf.
Food Standards Agency. *Nutrient Databank* web site. Available at: http://www.food.gov.uk/multimedia/spreadsheets/cofids.xls.
USDA, Agricultural Research Service. USDA National Nutrient Database for Standard Reference, Release 23. Available at: http://www.ars.usda.gov/ba/bhnrc/ndl.
Whitney, E.; Rolfes, S.R. *Understanding Nutrition*, 12th ed. Wadsworth: Belmont, CA, 2011, pp. 311–75.

Proteins and peptides:

Fulgoni, V.L. III; Keast, D.R.; Auestad, N.; Quann, E.E. Nutrients from dairy foods are difficult to replace in diets of Americans: Food pattern modeling and an analyses of the National Health and Nutrition Examination Survey 2003–2006. *Nutrition Research* 2011, *31*, 759–65.
Kwak, H.-S.; Ganesan, P.; Hong, Y.H. Nutritional benefits in cheese. In *Cheese: Types, Nutrition and Composition*; R.D. Foster, ed.; Nova Science Publishers: Hauppauge, NY, 2011, pp. 267–87.

Fat and cholesterol:

Bassett, C.M.C.; Edel, A.L.; Patenaude, A.F.; McCullough, R.S.; Blackwood, D.P.; Chouinard, P.Y.; Paquin, P.; Lamarche, B.; Pierce, G.N. Dietary vaccenic acid has antiatherogenic effects in LDLr$^{-/-}$ mice. *Journal of Nutrition* 2010, *140*, 18–24.
Chin, S.F.; Liu, W.; Storkson, J.M.; Ha, Y.L.; Pariza, M.W. Dietary sources of conjugated dienoic isomers of linoleic acid, a newly recognized class of anticarcinogens. *Journal of Food Composition and Analysis* 1992, *5*, 185–97.
Hjerpsted, J.; Leedo, E.; Tholstrup, T. Cheese intake in large amounts lowers LDL-cholesterol concentrations compared with butter intake of equal fat content. *American Journal of Clinical Nutrition* 2011, *94*, 1479–84.

Kim, J.H.; Kwan, O.-J.; Choi, N.J.; Oh, S.J.; Jeong, H.-Y.; Song, M.-K.; Jeong, I.; Kim, Y.J. Variations in conjugated linoleic acid (CLA) content of processed cheese by lactation time, feeding regimen, and ripening. *Journal of Agricultural and Food Chemistry* 2009, *57*, 3235–39.

Siri-Tarino, P.W.; Sun, Q.; Hu, F.B.; Krauss, R.M. Meta-analysis of prospective cohort studies evaluating the association of saturated fat with cardiovascular disease. *American Journal of Clinical Nutrition* 2010, *91*, 535–46.

Other health benefits:

Crichton, G.E.; Elias, M.F.; Dore, G.A.; Robbins, M.A. Relation between dairy food intake and cognitive function: The Maine-Syracuse Longitudinal Study. *International Dairy Journal* 2012, *22*, 15–23.

Elwood, P.C.; Pickering, J.C.; Givens, D.I.; Gallagher, J.E. The consumption of milk and dairy foods and the incidence of vascular disease and diabetes: An overview of the evidence. *Lipids* 2010, *45*, 925–39.

German, J.B.; Gibson, R.A.; Krauss, R.M.; Nestel, P.; Lamarche, B.; van Staveren, W.A.; Steijns, J.M.; de Groot, L.C.P.G.M.; Lock, A.L.; Destaillats, F. A reappraisal of the impact of dairy foods and milk fat on cardiovascular disease risk. *European Journal of Nutrition* 2009, *48*, 191–203.

Huth, P.J.; Park, K.M. Influence of dairy product and milk fat consumption on cardiovascular disease risk: A review of the evidence. *Advances in Nutrition* 2012, 3, 266–85.

Kashket, S.; DePaola, D.P. Cheese consumption and the development and progression of dental caries. *Nutrition Reviews* 2002, *60*, 97–103.

McGrane, M.M.; Essery, E.; Obbagy, J.; Lyon, J.; Mac Neil, P.; Spahn, J.; Van Horn, L. Dairy consumption, blood pressure, and risk of hypertension: An evidence-based review of recent literature. *Current Cardiovascular Risk Reports* 2011, *5*, 287–98.

Nagpal, R.; Behare, P.V.; Kumar, M.; Mohania, D.; Yadav, M.; Jain, S.; Menon, S.; Parkash, O.; Marotta, F.; Minelli, E.; Henry, C.J.K.; Yadav, H. Milk, milk products, and disease free health: An updated overview. *Critical Reviews in Food Science and Nutrition* 2012, *52*, 321–33.

O'Brien, N.M.; O'Connor, T.P. Nutritional aspects of cheese. *CCPM*, Vol. 1, pp. 573–81.

Zemel, M.B.; Thompson, W.; Milstead, A.; Morris, K.; Campbell, P. Calcium and dairy acceleration of weight and fat loss during energy restriction in obese adults. *Obesity Research* 2004, *12*, 582–90.

Why people buy cheese:

Moskowitz, H.; Reisner, M. Picking consumers' brains with Mind Genomics. *Food Technology* 2011, *65*(11), 24–31.

On the other hand:

McGee, H. *On Food and Cooking*. Scribner: New York, 2004, p. 58.

Neuman, W. F.D.A. and dairy industry spar over testing of milk. *New York Times*, January 25, 2011. Available at: http://www.nytimes.com/2011/01/26/business/26milk.html.

CHAPTER 15

Microbiology:

Beresford, T.P.; Fitzsimons, N.A.; Brennan, N.L.; Cogan, T.M. Recent advances in cheese microbiology. *International Dairy Journal* 2001, *11*, 259–74.

Structure:

Kalab, M. *Foods Under the Microscope* web site. Available at: http://www.magma.ca/~scimat/.
Rosenberg, M.; McCarthy, M.; Kauten, R. Evaluation of eye formation and structural quality of Swiss-type cheese by magnetic resonance imaging. *Journal of Dairy Science* 1992, *75*, 2083–91.

Texture:

Foegeding, E.A.; Brown, J.; Drake, M.A.; Daubert, C.R. Sensory and mechanical aspects of cheese texture. *International Dairy Journal* 2003, *13*, 585–91.

Color and melting:

Kindstedt, P.S.; Rippe, J.K. Rapid quantitative test for free oil (oiling off) in melted mozzarella cheese. *Journal of Dairy Science* 1990, *73*, 867–73.
Rohm, H.; Jaros, D. Colour of hard cheese. 1. Description of colour properties and effects of maturation. *Zeitschrift für Lebensmitteluntersuchung und Forschung A* 1996, *203*, 241–44.

Flavor comparisons:

Curioni, P.M.G.; Bosset, J.O. Key odorants in various cheese types as determined by gas chromatography-olfactometry. *International Dairy Journal* 2002, *12*, 959–84.
De Noni, I.; Battelli, G. Terpenes and fatty acid profiles of milk fat and "Bitto" cheese as affected by transhumance of cows on different mountain pastures. *Food Chemistry* 2008, *109*, 299–309.
Dumont, J.P.; Roger, S.; Cerf, P.; Adda, J. Etude de composés volatils neuters présents dans le Vacherin. *Le Lait* 1974, *54*, 243–51. [In French.]
Frank, D.C.; Owen, C.M.; Patterson, J. Solid phase microextraction (SPME) combined with gas-chromatography and olfactometry-mass spectrometry for characterization of cheese aroma compounds. *Lebensmittel-Wissenschaft und Technologie* 2004, *37*, 139–54.
Lecanu, L.; Ducruet, V.; Jouquand, C.; Gratadoux, J.J.; Feigenbaum, A. Optimization of headspace solid-phase microextraction (SPME) for the odor analysis of surface-ripened cheese. *Journal of Agricultural and Food Chemistry* 2002, *50*, 3810–17.

Lexicon:

Delahunty, C.M.; Drake, M.A. Sensory character of cheese and its evaluation. *CCPM*, Vol. 1, pp. 455–87.
Drake, M.A.; McIngvale, S.C.; Gerard, P.D.; Cadwallader, K.R.; Civille, G. Development of a descriptive language for Cheddar cheese. *Journal of Food Science* 2001, *66*, 1422–27.

Cheese contests:

Partridge, J.A. Cheddar and Cheddar-Type Cheese. In *The Sensory Evaluation of Dairy Products*; S. Clark, M. Costello, M.A. Drake, F. Bodyfelt, eds. Springer: New York, 2009, pp. 225–70.

Grading cheese:

USDA, Agricultural Marketing Service, Dairy Grading Branch. *DA Instruction 918-I. Instructions for Dairy Inspection and Grading Service.* October 1, 2009, edition. Available at: http://www.ams.usda.gov/AMSv1.0/getfile?dDocName=STELPRDC5069773.

USDA, Agricultural Marketing Service, Dairy Programs. *United States Standards for Grades of Swiss Cheese, Emmentaler Cheese.* February 22, 2001 edition. Available at: http://www.ams.usda.gov/AMSv1.0/getfile?dDocName=STELDEV3004468.

Cheese caper:

Kelly, S.; Heaton, K.; Hoogewerff, J. Tracing the geographical origin of food: The application of multi-element and multi-isotope analysis. *Trends in Food Science and Technology* 2005, *16*, 555–67.

Tunick, M.H.; Basch, J.J.; Maleeff, B.E.; Flanagan, J.F.; Holsinger, V.H. Characterization of natural and imitation Mozzarella cheeses by differential scanning calorimetry. *Journal of Dairy Science* 1989, *72*, 1976–80. Available at: http://wyndmoor.arserrc.gov/Page/1989\5395.pdf.

Imperial Russia and cheese science:

Gordin, M.D. *A Well-Ordered Thing: Dmitrii Mendeleev and the Shadow of the Periodic Table.* Basic Books: New York, 2004, p. 30.

Jensen, W.B. Letter to the Editor: Cheese or flu? *Bulletin of the History of Chemistry* 2007, *32*, 104.

Metchnikoff, É. *The Prolongation of Life: Optimistic Studies.* G.P. Putnam's Sons: New York, 1908, pp. 161–83.

Nobelprize.org. Ilya Mechnikov—Biography. Available at: http://www.nobelprize.org/nobel_prizes/medicine/laureates/1908/mechnikov.html.

Rabkina, N.A. *Great Soviet Encyclopedia*, 1979. Available at: http://encyclopedia2.thefreedictionary.com/Free+Economic+Society.

Vasiljevic, T.; Shah, N.P. Probiotics—From Metchnikoff to bioactives. *International Dairy Journal* 2008, *18*, 714–28.

CHAPTER 16

General:

Appropriate sections of *CLC, CP, 500*, and *CP*.

Code of Federal Regulations:

U.S. FDA. Cheeses and Related Cheese Products. *Code of Federal Regulations* Title 21, Pts. 133.138 and 133.142; 2011.

Xigalo Siteias:

Eur-Lex. Access to European Union Law. Available at: http://eur-lex.europa.eu/LexUriServ/LexUriServ.do?uri=OJ:C:2010:312:0025:0030:EN:PDF.

PDO, PGI, and TSG cheeses:

Engelmann, B.; Holler, P. *Gourmet's Guide to Cheese.* H.F. Ullmann: Königswinter, Germany, 2008.

Austria:

Lebensministerium. Traditionelle Lebensmittelle. Available at: http://www.traditionelle-lebensmittel.at/article/list/26084.

France:

Masui, K.; Yamada, T. *French Cheese.* DK Publishing: New York, 2005.

Greece:

Kitrilakis, S. The PDO cheeses of Greece. Available at: http://www.kerasma.gr/default.asp?entryID=407&pageID=96&tablePageID=35&langID=2.

Italy:

Formaggio.it. *Il Portale del Formaggio* web site. Available at: http://www.formaggio.it/home.htm [in Italian].

Portugal:

Smith, A. *The Ultimate User's Guide to Portuguese Cheese* web site. Available at: http://catavino.net/part-1-the-ultimate-user's-guide-to-portuguese-cheese.

Spain:

Profesionales del sector lácteo. *Queseros* web site. Available at: http://www.queseros.com [in Spanish].

United Kingdom:

British Cheese Board. British Protected Name Cheeses. Available at: http://www.britishcheese.com/othercheese/british_protected_name_cheeses-89.

Adulteration:

Accum, F. *A Treatise on Adulterations of Food, and Culinary Poisons.* J. Mallet: London, 1820, pp. 290–96.

Centre for Retail Research. The global retail theft barometer 2011. Available at: http://www.retailresearch.org/grtb_currentsurvey.php.

Tannahill, R. *Food in History*. Three Rivers Press: New York, 1988, p. 294.
Tunick, M.H. Dairy innovations over the past 100 years. *Journal of Agricultural and Food Chemistry* 2009, *57*, 8093–97. Available at: http://wyndmoor.arserrc.gov/Page/D8100\8127-Dairy100.pdf
Tunick, M.H. Approaches to detecting mislabelled cheeses. In *Cheese Yield and Factors Affecting Its Control*, Proceedings of the International Dairy Federation Seminar, Cork, Ireland, 1993 pp. 520–27. Available at: http://wyndmoor.arserrc.gov/Page/1994\6027.pdf.
USDA, Bureau of Chemistry. *Foods and Food Adulterants. Part First: Dairy Products*. U.S. Government Printing Office: Washington, DC, 1887, pp. 122–26.

U.S. cheese imports and exports:

USDA, Foreign Agricultural Service. Dairy monthly imports, July 2012. Available at: http://www.fas.usda.gov/itp/imports/monthly/2012/June/CIRJUNE 2012 DATA.pdf.
U.S. Dairy Export Council. U.S. exports of cheese. Available at: http://usdec.files.cms-plus.com/TradeData/PDFs/CHEESE VOL.pdf.

Cheese consumption around the world:

Canada Dairy Information Centre. Total and retail cheese consumption. Available at: http://www.dairyinfo.gc.ca/index_e.php?s1=dff-fcil&s2=cons&s3=consglo&page=tc-ft.

Do you want flies with that?:

Cockburn, A.; Barraco, R.A.; Reyman, T.A.; Peck, W.H. Autopsy of an Egyptian mummy. *Science* 1985, *187*, 1155–60.
Ente Regionale di Sviluppo e Assistenza Tecnica in Agricoltura. *Casu Frazigu-Formaggi* web site. Available at: http://www.sardegnaagricoltura.it/documenti/14_43_20070607153029.pdf. [In Italian.]
Glenday, C. *Guinness World Records 2009*. Bantam Books: New York, 2008.
Goff, M.L. *A Fly for the Prosecution: How Insect Evidence Helps Solve Crimes*. Harvard University Press: Cambridge, MA, 2001.

Mites:

Melnyk, J.P.; Smith, A.; Scott-Dupree, C.; Marcone, M.F.; Hill, A. Identification of cheese mite species inoculated on Mimolette and Milbenkase cheese through cryogenic scanning electron microscopy. *Journal of Dairy Science* 2010, *93*, 3461–68.
Peace, D.M. Reproductive success of the mite *Acarus Siro L.* on stored cheddar cheese of different ages. *Journal of Stored Product Research* 1983, *19*, 97–104.
Rorher, F. Cheese mites and other wonders. Available at: http://newsvote.bbc.co.uk/2/hi/uk_news/magazine/7423847.stm.

CHAPTER 17

Making Monterey Jack:

Karlin, M.; Anderson, E. *Artisan Cheese Making at Home: Techniques and Recipes for Mastering World-Class Cheeses*. Ten Speed Press: Berkley, CA, 2011, pp. 96–97.

Serving cheese:

500, 154–55.
CP, pp. 18–33.
Ensrud, B. *Wine with Food*. Fireside: New York, 1991, pp. 44–45.
Fletcher, J. *The Cheese Course: Enjoying the World's Best Cheeses at Your Table*. Chronicle Books: San Francisco, CA, 2000, pp. 9–17.
McCalman, M.; Gibbons, D. *Mastering Cheese: Lessons for Connoisseurship from a Maître Fromager*. Clarkson Potter: New York, 2009, pp. 204–17.

Other things to try:

Ackerman, S. *WxWise* web site. Available at: http://cimss.ssec.wisc.edu/wxwise/bluesky.html.
Gibbs, W.W.; Myhrvold, N. Microwaves and the speed of light. *Scientific American* 2011, *305*(12), 20.

Index

After cheese comes nothing.
— JAMES CLARKE, *Proverbs English and Latin*, 1639